MARE Publication Series

Volume 18

Series editors
Maarten Bavinck, University of Amsterdam, The Netherlands
J.M.Bavinck@uva.nl
Svein Jentoft, UiT-The Arctic University of Norway, Norway
Svein.Jentoft@uit.no

The MARE Publication Series is an initiative of the Centre for Maritime Research (MARE). MARE is an interdisciplinary social-science network devoted to studying the use and management of marine resources. It is based jointly at the University of Amsterdam and Wageningen University (www.marecentre.nl).

The MARE Publication Series addresses topics of contemporary relevance in the wide field of 'people and the sea'. It has a global scope and includes contributions from a wide range of social science disciplines as well as from applied sciences. Topics range from fisheries, to integrated management, coastal tourism, and environmental conservation. The series was previously hosted by Amsterdam University Press and joined Springer in 2011.

The MARE Publication Series is complemented by the Journal of Maritime Studies (MAST) and the biennial People and the Sea Conferences in Amsterdam.

More information about this series at http://www.springer.com/series/10413

Henrik Ringbom

Editor

Regulatory Gaps in Baltic Sea Governance

Selected Issues

 Springer

Editor
Henrik Ringbom
Department of Law
Åbo Akademi University
Turku/Åbo, Finland

ISSN 2212-6260   ISSN 2212-6279   (electronic)
MARE Publication Series
ISBN 978-3-319-75069-9   ISBN 978-3-319-75070-5   (eBook)
https://doi.org/10.1007/978-3-319-75070-5

Library of Congress Control Number: 2018935264

Printed on acid-free paper

This Springer imprint is published by the registered company Springer International Publishing AG part of Springer Nature.
The registered company address is: Gewerbestrasse 11, 6330 Cham, Switzerland

Series Foreword

The MARE Publication Series first paid attention to the complexities of the Baltic Sea and its human dimensions in 2016. The volume edited by Michael Gilek, Mikael Karlsson, Sebastian Linke and Katarzyna Smolarsz, entitled 'Environmental governance of the Baltic Sea' investigated a key set of environmental challenges and the ways in which they are currently being addressed. The present volume, in contrast, edited by Henrik Ringbom, examines the same regional sea, but takes a socio-legal perspective. The multiple layers of regulation that co-exist are the starting point of analysis, with attention subsequently turning to the gaps and uncertainties that emerge at their interstices. This focus on regulatory hotspots and their transmutations is instructive and makes for interesting analysis. We are therefore delighted to be able to include this volume in our series. As series editors, we also hope to publish similar in-depth analyses of other regional seas around the world in the future. Although each regional sea is expected to have unique features and challenges, it is also reasonable to assume that there are many similarities, which would allow for cross-regional learning, especially related to environmental, socio-political and legal governance.

The MARE Publication Series commenced in 2004 with Amsterdam University Press, but moved to Springer Academic Publishers in 2012. It has hitherto contained eighteen edited and single-authored volumes on a variety of regions and topics in the field of people, coasts and seas. Fritz Schmuhl and other staff of Springer have facilitated the production process, for which we are again more than grateful.

The series editors,
University of Tromsø, Tromsø, Norway   Svein Jentoft
University of Amsterdam, Amsterdam, Netherlands   Maarten Bavinck

Preface

This publication forms part of a research project on the regulatory 'anatomy' and governance structures of the Baltic Sea and for the Baltic Sea region. The BaltReg-project (2015–2018), run by Åbo Akademi University, is a joint interdisciplinary law and public administration research project, funded by the Academy of Finland, to analyse the interaction and interrelationship between the different layers of regulation in the region.

The interest in improving the understanding of the regulatory structures for the Baltic Sea has also been highlighted in the active cooperation between Åbo Akademi University and University of Turku, both located at the heart of the Baltic Sea in the city of Turku. Through the Baltic Sea Area Legal Studies (BALEX) network (www.balex.fi), the two universities and their partners around the region constantly seek to improve interdisciplinary understanding of how different regulatory measures affect the Baltic Sea.

Even if the Baltic Sea can arguably be described as the world's most heavily regulated sea area, there is surprisingly little analysis of how different regulatory layers interact and how the various governance regimes and institutions, laws and organisations that govern the area work together. This book addresses the legal interaction between various regulatory layers through the selection of a number of case studies on issues that are of particular relevance for the Baltic Sea. Later publications in the project will place more emphasis on the inter-relationship between law and other steering mechanisms.

The book represents the outcome of a small-scale international seminar, entitled 'Regulatory Voids and Legal Hotspots in the Baltic Sea', which was convened in Turku on 28–29 April 2016 and in which all authors participated. On behalf of the organisers of the event, I wish to extend my thanks to everybody who contributed. Particular thanks are also due to Springer Verlag for their helpfulness and support in securing the smooth publication of the volume and to the anonymous reviewers for their helpful and insightful comments. The seminar, as well as this book, are at the same time a celebration of Professor emeritus Peter Wetterstein's life-long work and devotion to maritime and environmental legal studies. Finally, thanks are owed to Finska Vetenskapssocieteten (the Finnish Society of Sciences and Letters) for their financial support for the seminar and to LL.M. Åsa Gustafsson for her excellent editorial assistance throughout the book project.

Turku/Åbo, Finland   Henrik Ringbom
20 January 2017

Contents

Chapter 1
Introduction

Henrik Ringbom

Abstract The introductory chapter explains the purpose and context of the book and briefly introduces the contributing articles.

Keywords Regulatory Gaps · Baltic Sea · Multi-level governance · Environmental regulation · Regulatory layers

The Baltic Sea region is unique in many ways, in terms of its geographical and climatic conditions and environmental challenges, but also in terms of its economic and political characteristics and governance structures. The area has undergone significant changes over recent decades, due to changing political landscapes and economic development in the region, as well as through the enlargement and increased activities of the European Union.

The focus of this publication is the uniqueness of the Baltic Sea from a legal perspective. Up to six layers of regulation (general international law, regional conventions, EU law, national laws, local and municipal rules plus a whole range of non-binding norms and other 'soft law' arrangements) act in parallel in the region. However, a large number of regulatory layers does not in itself ensure consistency or effectiveness. When the regulatory landscape is approached from the point of view of individual substantive topics, it is apparent that the norms of different regulatory layers entail both overlaps, gaps and uncertainties. The rules of different layers are inter-related through a complex and constantly evolving relationships, which vary from one subject area to another, thus needing to be assessed case-by-case.

This book focuses on certain gaps or other legal 'hotspots' in the Baltic Sea region. The individual chapters study issues that are deemed to be particularly topical from a Baltic regional perspective, addressing maritime issues that are decidedly international in scope, yet entail legal uncertainties at international or domestic

H. Ringbom (✉)
Department of Law, Åbo Akademi University, Turku/Åbo, Finland
e-mail: henrik.ringbom@abo.fi

© Springer International Publishing AG, part of Springer Nature 2018
H. Ringbom (ed.), *Regulatory Gaps in Baltic Sea Governance*, MARE
Publication Series 18, https://doi.org/10.1007/978-3-319-75070-5_1

1

level. The texts represent legal analyses in their own right, covering a broad spectrum of public and private law issues as they are addressed in international, EU and national laws. In terms of substance, the issues range from a geographic review of the key regulatory gaps and topics in the Baltic Sea area (Chaps. 2 and 3) to a closer review of issues that are decidedly international in nature - and highly relevant for the Baltic Sea region - yet not comprehensively regulated at the international or EU level (Chaps. 4, 5, 6, 7, 8, 9 and 10). The selected issues represent different starting points in terms of how well the topic is regulated and which regulatory layer dominates.

In addition to presenting a set of legal analyses of topical issues for the region, which in itself is a meritorious objective in view of the relative scarcity of legal studies about the Baltic Sea, the publication also seeks to study the regulatory 'anatomy' of the selected issues in some more detail. Through the legal analyses the chapters explore how regulatory gaps are formed, how they are filled, how the rules of the different layers work together and interact with each other in the selected areas. Accordingly, the secondary ambition is to explore, through the chapters, whether more general conclusions can be drawn about the nature of the regulatory gaps and multi-layerism in order to produce a better understanding of how regulations on multiple levels operate in practice.

To begin with, the Baltic Sea is one of the most complete sea areas in the world in terms of maritime delimitation. In his chapter, Erik Franckx illustrates that, apart from a few issues which are still to be settled, the whole sea area is now delimited. Rights and duties relating to different uses of the sea and its resources are thus distributed between the coastal states through a range of international agreements signed by the coastal states concerned. The 'high seas' areas of the Baltic Sea have disappeared in the process along with any other area beyond national jurisdiction. There are no more 'no man's lands' in the Baltic Sea or its seabed, which strengthens the picture - and jurisdictional reality - that questions related to the regulation and usage of the Baltic Sea and its resources are now for the Baltic Sea littoral states themselves to regulate and resolve.

Nevertheless, settling the boundaries of the maritime zones does not resolve all issues relating to the maritime zones. As the chapter by Pirjo Kleemola-Juntunen demonstrates, fundamental questions relating to the rights and usage of marine areas remain, even in the Baltic Sea. Through her analysis of the Baltic Sea's international straits (i.e. the Danish Straits and the Strait of Åland) she illustrates how fundamental questions related to the rights and obligations of passage are still disputed. The very nature of the straits in question is still not agreed on by all the relevant states. In this case, Dr. Kleemola-Juntunen finds that part of the answer to establishing the true legal nature of the states lies in history and that it is necessary to go back to the pre-UNCLOS sources to establish the legal nature of the straits.

What is commonly regarded as the most serious environmental threat to the Baltic Sea is the excessive influx of nutrients into the sea, or eutrophication. Despite widespread acknowledgment of the crucial correlation between emissions of nutrients into the Baltic Sea and the health of the sea, there is relatively little clear-cut regulation for such emissions. Eutrophication is not a concern at global level and there are no global regulatory instruments to deal with this issue. By contrast, the topic has received considerable interest within the Helsinki Commission, HELCOM, which is specifically tasked to deal with the environmental challenges of the Baltic Sea and the EU. The way this matter is approached in regulation illustrates several complexities in the more modern eco-system and goal-oriented way of addressing environmental threats. As Brita Bohman shows in her chapter, on the one hand, the rules that follow from the Helsinki Convention are relatively unspecific. Many of the concrete targets are laid down in the Baltic Sea Action Plan, the legal status of which is not entirely clear. On the other hand, EU regulation in this area is increasingly eco-system-based, striving for generic goals such as a 'good environmental status' and hence allowing significant liberties to set the goals and measures region by region. Bohman illustrates the complexities in this regulatory situation, which is marked by a series of principles, targets and cross-references, where it is often unclear which body - if any - has a mandate to translate the target-oriented goals to more enforceable obligations and to follow up their implementation.

The second most important category of environmental threat to the Baltic Sea relates to pollution from various forms of chemicals. In this area, the present book analyses the regulatory gap that relates to chemical mixtures. Existing legislation on chemicals – even modern variants, such as the EU's REACH Regulation – is largely focused on individual substances, one-by-one, where the risk of the substance is based on its effect as the only toxic substance in an otherwise pristine environment. This focus fails to target combined toxicity when different chemicals are mixed, creating different types of regulatory gaps and imperfections. For example, the toxicity of a mixture of chemicals may very well exceed the toxicity of each individual compound and small, individually non-toxic concentrations might equally well add up to create severe toxicity of the overall 'cocktail'. The extent and significance of this regulatory gap, as well as the legal efforts to manage chemical mixes, are discussed in the chapter written by Lena Gipperth.

The remaining substantive issues considered in the book relate to different aspects of the seabed in the Baltic Sea. While the main jurisdictional rights and obligations relating to the seabed are laid down in UNCLOS, our examples illustrate that those provisions alone are rarely sufficient to address concrete issues. Accordingly, many of the topics addressed in the book have been complemented by more specific international rules at global, regional or EU-level, but the additional layers of laws have not always contributed to greater regulatory clarity.

One example is the regulation of historic shipwrecks, many of which are remarkably well preserved in the Baltic Sea, hence justifying a specific regional attention to this topic. Jan Aminoff explores the regulatory situation in his article and notes that in the absence of any widespread ratification of the 2001 UNESCO Convention on historical wrecks, there are significant gaps and uncertainties in the international regulation of wrecks. Thematically wrecks fall into a cross-section between the law of the sea, salvage law, public and private law, which caters for a variety of solutions to fill such legal gaps at national and Nordic level.

A different aspect of wrecks is addressed in the chapter by Markku Suksi, where he assesses the rights and obligations of public authorities to take action against wrecks and cargo. Until recently this matter has been subject to important regulatory voids and uncertainty at both international and domestic levels, but several recent and important developments have sought to close or at least reduce those gaps. Even following such amendments, the Finnish legislation on wrecks contains a variety of acts, authorities and alternative procedural bases for the actions of authorities. Gaps have not been entirely removed, but at least reduced in scope, thanks to guidance by international rules. Significant variations still exist between national legislations in this area, not least among EU member states, as the matter has not been subject to regulation at the EU-level.

The regulation of subsea pipelines is explored in the chapter by Peter Wetterstein which addresses various questions related to civil liability and compensation departing from the Nord Stream gas pipeline that traverses the Baltic Sea. For such pipelines, too, UNCLOS provides the overall jurisdictional framework, by ensuring that states have significant rights to lay pipelines on the exclusive economic zones and continental shelves of other states, subject to certain obligations. For the rest of the issues, there is no international legal framework in place for this type of projects. Despite its inherently transnational nature, the construction of Nord Stream is based on a series of bilateral negotiations and agreements between the operator and the coastal (shelf) state concerned, including environmental impact assessments under the Espoo Convention. The absence of broad regulation for the Baltic Sea on this issue means that not only the process relating to the construction of a pipeline, including permits and relevant criteria, will vary from one state to another, but also that the subsequent legal status of the pipeline and related risks differ from one state to another. In his chapter, Peter Wetterstein illustrates how these divergences in national laws affect the application of civil liability in the case of damage caused by such pipelines.

The final substantive chapter concerns carbon capture and storage under the seabed. This represents an example of a field which – despite its recent appearance – is subject to a wealth of international and regional rules. In contrast to other areas discussed above, there is currently not much practical experience regarding this in the Baltic Sea. Accordingly, the chapter by David Langlet approaches the matter in tentative terms, considering the extent to which such activity is permitted on the Baltic seabed, and assessing the different tools available for prioritising between the different and conflicting uses. This leads to more recent environmental legislation, notably at EU-level, which focuses on processes and procedures, including marine spatial planning, and thereby more general questions relating to multi-level regulation of the Baltic Sea.

The selection of topics provides an interesting sample of different regulatory starting points for dealing with issues that are of particular relevance to the Baltic Sea. The final chapter provides a brief summary of the findings in terms of governance and assesses whether more general conclusions can be drawn from the material. Henrik Ringbom and Marko Joas make certain general observations related to the interaction between various kinds (and layers) of laws. In addition, the concluding section briefly addresses the broader question of how other forms of (multi-level) governance structures find the space to operate within and between the existing (international, regional and national) rules. As it is probably not controversial to assume - as a starting point - that gaps and uncertainty in regulation increases operating space for other (non-legal) policy tools and steering mechanisms to influence the behaviour of states, sub-national governments and individuals, the findings on regulatory gaps will be of significance for analysing the interaction between law and other policy instruments in Baltic Sea governance in the later stages of the BaltReg project.

Chapter 2
Gaps in Baltic Sea Maritime Boundaries

Erik Franckx

Abstract Does the submission that the Baltic Sea is the world's most regulated international marine area also apply to maritime boundary delimitations? Probably so, according to this chapter, which addresses existing and past boundary agreements in the Baltic Sea. Following a general review of the law applicable to maritime boundary delimitation, it is concluded that even if the Baltic Sea is already fully covered by coastal zones, and that the areas of high seas or deep seabed have thus disappeared, there are still some outstanding issues and overlapping claims. Nevertheless, such a degree of completeness is unique in international comparison and, what is more, all boundary agreements in the Baltic Sea have been settled by negotiations, outside courts and tribunals.

Keywords Maritime delimitation · Law of the sea · Baltic Sea · Border agreements

2.1 Introduction

The present article is the reflection of an oral presentation given at an international seminar organized by the Åbo Akademi University, entitled "Regulatory Voids and Legal Hotspots in the Baltic Sea" held at Turku, Finland, 28–29 April 2016. Even though the purpose of this seminar was to analyze the interaction between the different layers of legal regulation applicable to the Baltic Sea, i.e. international, regional, European Union, national, and local levels, the present contribution will mainly address the first level (international) and to a lesser extent the fourth level (national legislation). As the determination of maritime boundaries are normally unilateral acts undertaken by the coastal states, the impression might be created that

E. Franckx (✉)
Vrije Universiteit Brussel, Brussels, Belgium
e-mail: erik.franckx@vub.be

© Springer International Publishing AG, part of Springer Nature 2018
H. Ringbom (ed.), *Regulatory Gaps in Baltic Sea Governance*, MARE
Publication Series 18, https://doi.org/10.1007/978-3-319-75070-5_2

7

the national level prevails in this particular domain. However, as clearly stated by the International Court of Justice in 1951:

> The delimitation of sea areas has always an international aspect; it cannot be dependent merely upon the will of the coastal State as expressed in its municipal law. Although it is true that the act of delimitation is necessarily a unilateral act, because only the coastal State is competent to undertake it, the validity of the delimitation with regard to other States depends upon international law.[1]

It is in other words the international level that the present contribution will focus upon. It will consequently also be on this particular level that the regulatory voids will be looked for. The paper intends to find out whether the generally accepted submission that the Baltic Sea is probably the world's most regulated international marine area, also applies to maritime boundary delimitations in the area. Within international law, the inner concentric circles of relevance here are the law of the sea, maritime delimitation law and finally the Baltic Sea as the latter constitutes the geographical field of application covered by the present publication.

Four sections will be subsequently addressed. Firstly, the broader international legal framework will be highlighted, including the applicable primary sources and the legal principles to be found there concerning maritime delimitation. Secondly, the focus will shift towards the Baltic Sea and the way these general legal principles have been applied in practice there. Thirdly, the remaining gaps will be uncovered, which will allow the paper to finally assess where exactly the Baltic Sea stands at present compared to other marine areas around the globe.

2.2   Applicable Primary Sources and Legal Principles

Contrary to some other branches of international law, the international law of the sea stands out for being well-codified.[2] This did not come easy, however, as a partial attempt undertaken by the League of Nations utterly failed during the 1930.[3] For the United Nations, however, the codification of the law of the sea has been a major success story. This organization was not only able to codify this particular branch of

---

[1] *Fisheries Case*, Judgment of 18 December 1951 (1951) International Court of Justice (ICJ) Reports 1951, 116, 132.

[2] T. Treves "Law of the Sea" in R. Wolfrum (ed) *Max Planck Encyclopedia of Public International Law Online* (Oxford, Oxford University Press, 2011) paras. 11–21 (available at <www.mpepil. com>), who uses the following title above these paragraphs: "The Law of the Sea as a Codified Branch of International Law."

[3] It concerned only the regime of the territorial waters, but mainly due to the divergent opinions that existed at that time concerning the breadth of that maritime zone, the conference failed to adopt a convention on this subject. For an authoritative account of the law of the sea as it existed at that time, see the three volumes of G.C. Gidel *Le droit international public de la mer: le temps de paix* (Mellottée, Chateauroux, 1932–1934).

international law a first time in 1958,[4] but it did so a second time in 1982,[5] producing a single document which today is generally referred to as the Constitution for the Oceans,[6] as envisaged by its drafters in 1982.[7]

The first place to look for legal rules governing maritime delimitation is consequently the 1958 conventional system as well as the UNCLOS. Three types of delimitation are involved when considering the different maritime zones codified in 1958 and 1982: The first concerns the starting point for measuring these zones, i.e. the baseline; the second relates to their outer limit; and the third, finally, concerns the eventuality that these maritime entitlements of adjacent or opposite states overlap.[8] Only the third type of delimitation just enumerated will be addressed here. Since there are nine coastal states in the Baltic Sea, as defined by the World Hydrographic Organization,[9] and because its width is nowhere more than 400 nautical miles between countries having to delimit their maritime zones *inter se*, this entails that a good number of maritime boundaries need to be delimited.[10]

It is important to note that the rules to be found in the 1958 conventional system are not totally identical to those included in the UNCLOS. While there is no difference in substance between art. 12 of the 1958 Convention on the Territorial Sea and the Contiguous Zone and art. 15 of the UNCLOS,[11] both governing the delimitation

---

[4] Four conventions were adopted at that time: Convention on the Territorial Sea and the Contiguous Zone of 29 April 1958 (1966) 516 *United Nations Treaty Series* (UNTS) 205, 206–22; Convention on the Continental Shelf of 29 April 1958 (1965) 499 UNTS 311, 312–320; Convention on the High Seas of 29 April 1958 (1964) 450 UNTS 11, 82–102; and Convention on Fishing and Conservation of the Living Resources of the High Seas of 29 April 1958 (1967) 559 UNTS 285, 286–300 all available at <https://treaties.un.org/doc/Treaties/1966/03/19660320%20 02-16%20AM/Ch_XXI_01_2_3_4_5p.pdf>. Hereinafter 1958 conventional system.

[5] United Nations Convention on the Law of the Sea of 10 December 1982 (1998) 1833 UNTS 3, 397–581 available at <www.un.org/Depts/los/convention_agreements/texts/unclos/unclos_e.pdf>. Hereinafter UNCLOS.

[6] See remarks of the Secretary-General of the United Nations, Ban Ki-moon, made at the occasion of the commemoration of the thirtieth anniversary of the opening for signature of the UNCLOS, on 10 December 2012, before the General Assembly available at <http://www.un.org/press/en/2012/sgsm14710.doc.htm>.

[7] Remarks by Tommy T.B. Koh of Singapore, President of the Third United Nations Conference on the Law of the Sea, on 10 December 1982 available at <https://www.un.org/Depts/los/convention_agreements/texts/koh_english.pdf>.

[8] As noted in K. Gustafson Juras, J.E. Noyes and E. Franckx *Law of the Sea in a Nutshell* 2nd (West Publishing Company, St. Paul, Minnesota, 2010) 97.

[9] International Hydrographic Organization *Limits of Oceans and Seas (Special Publication N° 28)* 3rd (Imp. Monégasque, Monte-Carlo, 1953) 4–5 available at <http://www.iho.int/iho_pubs/standard/S-23/S23_1953.pdf>. Norway is thus not included for present purposes.

[10] The Russian enclave of Kaliningrad adds to its complexity.

[11] UNCLOS, note 5 at art. 15 reads: "Where the coasts of two States are opposite or adjacent to each other, neither of the two States is entitled, failing agreement between them to the contrary, to extend its territorial sea beyond the median line every point of which is equidistant from the nearest points on the baselines from which the breadth of the territorial seas of each of the two States is measured. The above provision does not apply, however, where it is necessary by reason of historic title or other special circumstances to delimit the territorial seas of the two States in a way which

of the territorial sea, this is not the case for the respective provisions relating to the delimitation of the continental shelf.

The 1958 Convention on the Continental Shelf contains a provision that closely resembles the one on the delimitation of the territorial sea, even though a difference is made this time between opposite and adjacent states,[12] the sole difference as to the substance of the rule being that historic title is no longer explicitly mentioned as a special circumstance that can offset the application of the median or equidistant line.[13] The UNCLOS has however "de-codified" the delimitation rule concerning the continental shelf,[14] because any concrete guidance as to the method to be applied has been eliminated in favor of a provision that only requires an equitable solution to be achieved.[15] Seminal in this development was the decision of the ICJ in the *North Sea Continental Shelf* cases of 1969, clearly indicating not only that art. 6 of the 1958 Convention on the Continental Shelf did not codify existing international law,[16] but also that this provision had not resulted in the creation of a new norm of customary international law since its codification,[17] which the ICJ later accepted with respect to the provision concerning the delimitation of the territorial sea as included in the 1958 Convention on the Territorial Sea and the Contiguous Zone and the UNCLOS in virtually identical terms.[18]

---

is at variance therewith." Only minor drafting changes are to be noted when this article is compared to the corresponding article of the 1958 Convention on the Territorial Sea and Contiguous Zone, note 4 at art. 12.

[12] The median line applies between opposite states, whereas equidistance governs the delimitation between adjacent states.

[13] 1958 Convention on the Continental Shelf, note 4 at art. 6 reads: "1) Where the same continental shelf is adjacent to the territories of two or more States whose coasts are opposite each other, the boundary of the continental shelf appertaining to such States shall be determined by agreement between them. In the absence of agreement, and unless another boundary line is justified by special circumstances, the boundary is the median line, every point of which is equidistant from the nearest points of the baselines from which the breadth of the territorial sea of each State is measured. 2) Where the same continental shelf is adjacent to the territories of two adjacent States, the boundary of the continental shelf shall be determined by agreement between them. In the absence of agreement, and unless another boundary line is justified by special circumstances, the boundary shall be determined by application of the principle of equidistance from the nearest points of the baselines from which the breadth of the territorial sea of each State is measured."

[14] The term "décodification" was coined in this respect by T. Treves "Codification du droit international et pratique des États dans le droit de la mer" (1990) 223 *Recueil des cours de l'Académie de droit international de la Haye* 9, 104.

[15] UNCLOS, note 5 at art. 83 reads: "The delimitation of the continental shelf between States with opposite or adjacent coasts shall be effected by agreement on the basis of international law, as referred to in art. 38 of the Statute of the International Court of Justice, in order to achieve an equitable solution."

[16] *North Sea Continental Shelf Cases*, Judgment of 20 February 1969 (1969) ICJ Reports 3, 36–41 paras. 60–69.

[17] Ibid., 41–45 paras. 70–81.

[18] *Case Concerning Maritime Delimitation and Territorial Questions Between Qatar and Bahrein*, Merits, Judgment of 16 March 2001 (2001) ICJ Reports 40, 94 para. 176. This finding of the ICJ seems therefore to drastically reduce the persuasive force of the remarks made by a country like

In accordance with the UNCLOS, the delimitation provision of the newly created exclusive economic zone simply follows the one on the continental shelf in that parties need to arrive at an equitable solution.[19] The fact that both the continental shelf and the exclusive economic zone extend to a minimum of 200 nautical miles under the UNCLOS, further explains why a different delimitation rule than the one applicable to the territorial sea, limited to a maximum of 12 nautical miles, proved sensible to the UNCLOS III negotiators.[20]

If the delimitation provisions of the territorial sea and continental shelf under the 1958 conventional system in other words had much in common, namely the application of the equidistance/special circumstances principle,[21] they became totally detached under the UNCLOS. The ICJ noted in the *North Sea Continental Shelf* cases that a difference existed between the *in casu* non-applicable conventional delimitation norm with respect to the continental shelf, incorporating the equidistance/special circumstances principle, and the corresponding rule of customary international law, rather emphasizing the equitable principles/relevant circumstances approach.[22] As the *North Sea Continental Shelf* cases clearly indicated that equidistance does not always lead to an equitable result, especially in the presence of convex and concave coastlines, it should not come as a surprise that during the negotiations of the third United Nations Conference on the Law of the Sea (UNCLOS III), governed by the rule of consensus, adherents of both approaches finally settled for the lowest common denominator, i.e. a formula in which any explicit reference to controversial notions such as "equidistance", "equitable principles", "special circumstance" and "relevant circumstances" was carefully avoided.[23]

As a consequence, the exact relationship between the 1958 conventional system and the UNCLOS deserves some attention in this respect especially as codified rules do not necessarily reflect customary international law. According to the UNCLOS, the latter document prevails between states parties over the relevant instruments of the 1958 conventional system.[24] If the countries involved in a maritime delimitation are states parties to the relevant instrument of the 1958 conventional system but at

---

Belgium in its declaration when signing the UNCLOS, stating that "it regrets that the concept of equity, adopted for the delimitation of the continental shelf and the exclusive economic zone, was not applied again in the provisions for delimiting the territorial sea."

[19] UNCLOS, note 5 at art. 74.

[20] D.R. Rothwell and T. Stephens *The International Law of the Sea* 2nd (Hart Publishing, Oxford, 2016) 421.

[21] S. Yanai "International Law Concerning Maritime Boundary Delimitation" in D.J. Attard, M. Fitzmaurice and N.A.M. Gutiérrez (eds) *The IMLI Manual on International Maritime Law, Volume I, The Law of the Sea* (Oxford University Press, Oxford, 2014) 304, 306–307.

[22] *North Sea Continental Shelf Cases*, note 16 at 53 para. 101, where the ICJ states: "[D]elimitation is to be effected by agreement in accordance with equitable principles, and taking account of all the relevant circumstances."

[23] As stressed by M.D. Evans "Maritime Boundary Delimitation" in D.R. Rothwell, A.G.O. Elferink, K.N. Scott and T. Stephens (eds) *The Oxford Handbook of the Law of the Sea* (Oxford University Press, Oxford, 2015) 254, 258.

[24] UNCLOS, note 5 at art. 311(1).

least one of them is not a party to the UNCLOS, art. 6 of the Convention on the Continental Shelf might in theory still be applicable between them even though the content of this article may not necessarily correspond with customary international law on the subject.[25] As all nine coastal states in the Baltic Sea ratified the UNCLOS between 1994[26] and 2005,[27] they are all bound as a matter of treaty law by arts. 15, 74 and 83 of that document.

Given the absence of any concrete guidance as to the method to be followed in arts. 74 and 83, the reference to an equitable solution appears to direct the countries with different views on the equitableness of the matter to third party settlement. Much more than through the practice of states–not bound to base their delimitation agreements on law–, the law on maritime delimitation has been mainly shaped through the decisions of courts and tribunals as a kind of judge-made common law.[28] Totally in line with the general trend, moreover, only very few Baltic Sea coastal states have excluded maritime delimitation from compulsory third party settlement as permitted under art. 298 (1)(a)(i) of the UNCLOS.[29]

One could therefore have expected that courts and tribunals played a major role in delimiting the maritime areas of the Baltic Sea, especially when taking into consideration the presence of many islands in the area, usually rendering the achievement of an equitable solution quite elusive.

The next part will nevertheless demonstrate that despite the absence of any concrete guidance with respect to the rules of delimitation concerning the continental shelf and the exclusive economic zone under the UNCLOS, not a single segment of maritime delimitation has so far been arrived at by means of third party settlement in the Baltic Sea. Instead, states have always succeeded in finding a solution through the conclusion of bi- and trilateral agreements up to the present.

---

[25] The term "in theory" is used, because there are in reality but a few countries bound by the 1958 Convention on the Continental Shelf today, that are not at the same time also a party to the UNCLOS. It concerns the following five countries: Cambodia, Colombia, Israel, United States and Venezuela. None of these states border the Baltic Sea.

[26] Germany. This country needed to ratify the UNCLOS before its entry into force if it wanted the seat of the International Tribunal for the Law of the Sea to be located in Hamburg.

[27] Estonia was the last Baltic Sea coastal state to ratify the UNCLOS on 16 August 2005.

[28] J.I. Charney "Progress in International Maritime Boundary Delimitation Law" (1994) 88 *American Journal of International Law* 227, 228.

[29] It concerns Denmark and Russia, who both did so at the time of ratification. Information available at <http://www.un.org/Depts/los/convention_agreements/convention_declarations. htm#Denmark%20Upon%20ratification>.

2.3   Existing Maritime Delimitation Agreements in the Baltic Sea[30]

All maritime boundaries in the Baltic Sea have so far been established by means of agreements directly arrived at between the coastal states in the area. When viewed as a whole,[31] four distinct periods can be distinguished in this respect.

Period 1: 1945–1972
This first period covers the heyday of the Cold War in Europe. This left a clear imprint on the nature of the bilateral delimitation agreements concluded during this period, as the majority of them were concluded between Eastern Bloc countries. The first, concluded in 1958, related to the delimitation of the territorial sea between Poland and the former Soviet Union.[32] The last, concluded between the same two parties in 1969, added a continental shelf segment to this boundary.[33] The political advantage of the conclusion of these agreements was partly the early acceptance in a treaty of "territorial waters" of 12 nautical miles, as claimed by the former Soviet Union at that time.[34] Also the treaty concluded between the former German Democratic Republic and Poland on the delimitation of the continental shelf in 1968[35] is noteworthy here because it implied that the former German Democratic

---

[30] This part is based on E. Franckx "Maritime Delimitation in the Baltic Sea: What Has Already Been Accomplished?" (2012) 6 (issue 3) *TransNav (International Journal on Marine Navigation and Safety of Sea Transportation)* 437–442 available at <http://www.transnav.eu/Article_Maritime_Delimitation_in_Franckx,23,382.html> and the more than 30 further references to be found in that article relating to maritime delimitation in the Baltic Sea written by the present author (ibid., 441–442).

[31] The present author has served as a regional expert for the Baltic Sea within the framework of a project set up by the American Society of International Law during the late 1980s, sponsored by the Ford and Mellon Foundations, which intended to provide an in-depth examination of the state practice arising from more than 100 existing ocean boundary delimitations. Two meetings, gathering all participants, were organized in order to outline and subsequently discuss the results of the project. A first one was held at Washington, D.C., 13–14 December 1988. The second one took place at Airlie, Virginia, 13–16 December 1989. Once the book, entitled International Maritime Boundaries, was published in 1993 (Volumes I and II), it was decided to prepare supplements at regular intervals. Volume III appeared in 1998, Volume IV in 2002, Volume V in 2005, Volume VI in 2011, and Volume VII in 2016. This part of the project is still running at present. In 1997, moreover, a CD-ROM version of this book was released. On 7 April 1994 the Certificate of Merit in the category of "high technical craftsmanship and high utility to practising lawyers and scholars" was attributed by the Executive Council of the American Society of International Law to this book.

[32] Protocol Concerning the Delimitation of Polish and Soviet Territorial Waters in the Gulf of Gdansk of the Baltic Sea of 18 March 1958 (1959) 340 UNTS 89, 94–96.

[33] Treaty on the Continental Shelf in the Gulf of Gdansk and the Southeastern Part of the Baltic Sea of 28 August 1969 (1971) 769 UNTS 75, 82–86.

[34] Indeed, the above-mentioned 1958 Protocol (note 32) was special in that it provided different terminal points for Poland, located three nautical miles seaward from the terminal point of the Polish-Soviet land boundary, and the former Soviet Union, located 12 nautical miles from that same starting point.

[35] Treaty Concerning the Delimitation of the Continental Shelf in the Baltic Sea of 29 October 1968 (1971) 768 UNTS 253, 260–264.

Republic, a country which had not been allowed to participate in the negotiations leading up to the 1958 conventional system or to sign up to any of the ensuing conventions, was nevertheless able to claim a continental shelf of its own, something which was very much contested at that time by its Western neighbour. The only non-Eastern Bloc country participating in these early delimitation efforts by Eastern Bloc countries was Finland, claiming strict neutrality in the East-West divide.

Period 2: 1973–1985

This second period starts after the normalization of relations between the two Germanies in 1972. This made it possible for these two countries, for the first time since the end of the Second World War, to establish normal treaty relations between themselves. It did not take long before they applied this novelty to their offshore areas as a bilateral treaty was concluded between them in 1974 dividing the waters of Lübeck Bay.[36] It was during 1973 that both Germanies were moreover admitted to the United Nations. This ushered in a totally new era for maritime delimitation agreements in the Baltic Sea as countries from the Eastern and Western Bloc started to conclude such agreements among themselves. The first such agreement concerning the delimitation of the continental shelf was concluded between the former German Democratic Republic and Sweden in 1978.[37] But no matter how important these agreements were from a political point of view, their practical significance from a maritime delimitation perspective remained rather limited.[38]

Period 3: 1985–1990

Contrary to the previous period, here it is rather the end year that was determined by the occurrence of an important political event in the area, namely the dissolution of the former Soviet Union. This may have been the shortest of the three periods considered so far, but at the same time it should be emphasized that it has been a most productive one. In half a decade more delimitation agreements were concluded than during the preceding forty years.

As coastal states normally address the more difficult areas at a later stage, this period is also the most interesting one from a maritime delimitation point of view. A good example of such a delimitation process burdened by the presence of sizeable islands close to the median line can be found in the situation that existed between the former Soviet Union and Sweden because of the particular location of the Swedish islands of Gotland and Gotska Sandön in the middle of the area to be delimited. In such situations coastal states often have opposing views on the effect to be attributed to such islands: The country in possession of such islands wants

---

[36] Protocol Note Concerning the Boundary in Lübeck Bay of 29 June 1974, reprinted in English translation in (1976) 74 *Limits in the Seas* 1.

[37] Agreement about the Delimitation of the Continental Shelf of 22 June 1978 (1979) 1147 UNTS 193, 198–199.

[38] Indeed, the above-mentioned 1974 Protocol (note 36) only concerned a boundary of about 8 nautical miles, whereas the 1978 Agreement (note 37) concerned an area devoid of special features like islands and also measured only about 29 nautical miles.

them to generate full effect, whereas the other country rather wishes to construct the median line starting from the mainland coastlines, i.e. giving no effect to the islands in question. This was also the starting point of the negotiations between the former Soviet Union and Sweden, which lasted for about 20 years. The compromise finally reached in 1988 was that the disputed area, calculated on the basis of the two different starting positions of the parties, would be divided by attributing 75% to Sweden and 25% to the former Soviet Union. In the zones thus created, the other party received fishing rights in reverse percentages, meaning that the former Soviet Union received 75% fishing rights in the Swedish zone, whereas Sweden only received 25% fishing rights in the Soviet zone.[39] The maritime delimitation between Poland and Sweden arrived at a year later was confronted with a similar problem and received a similar solution.[40]

It was also during this period that coastal states in the Baltic Sea started to address the new maritime zone created by the UNCLOS III negotiations of interest to this region, namely the exclusive economic zone.[41] Even though the UNCLOS only entered into force on 16 November 1994, the notion of the exclusive economic zone had already assertively been established by the ICJ as forming part of customary international law by the beginning of this third period.[42] Not all countries, however, had established an exclusive economic zone by the end of this period, so some ingeniousness needed to be displayed to also suit those countries that still had only created fishery zones.[43]

This was finally also the period that the first trilateral agreement was concluded in the Baltic Sea governing the fixing of a tripoint between the three countries involved, namely Poland, Sweden and the former Soviet Union.[44]

Period 4: 1991-Present
This is the last period to be mentioned, as it is still running. As already stated, it was triggered by the disappearance of the former Soviet Union from the political map of the world during the month of December 1991. The influence of this event on maritime delimitation in the Baltic Sea was considerable. Three countries re-emerged as coastal states in the Baltic Sea, namely Estonia, Latvia and Lithuania, while one had

---

[39] Agreement on the Delimitation of the Continental Shelf and of the Swedish Fishing Zone and the Soviet Economic Zone in the Baltic of 18 April 1988 (2001) 1557 UNTS 275, 283–285.

[40] Agreement on the Delimitation of the Continental Shelf and the Fishery Zones of 10 February 1989 (1999) 1590 UNTS 361, 365–366.

[41] Because the Baltic Sea is totally covered by exclusive economic zones of coastal states, the second new zone created by these UNCLOS III negotiations, i.e. the Area, is of no particular interest to the region.

[42] Even though the ICJ had already hinted a few times at this possibility, it was only in 1985 that this Court firmly stated that the exclusive economic zone did form part of customary international law. About this gradual development, see Gustafson Juras, Noyes and Franckx, note 8 at 256–258.

[43] The 1988 Agreement (note 39) is a good example as indicated by its title.

[44] Agreement Concerning the Junction Point of the Maritime Boundaries in the Baltic Sea of 30 June 1989 (1999) 1590 UNTS 299, 306–307.

disappeared shortly before, namely the former German Democratic Republic after the German reunification in 1990. The former East-West divide in the area has moreover taken on a totally new dimension because all Baltic Sea coastal states had joined the European Union by 2004, with the sole exception of the Russian Federation.

From a maritime delimitation point of view this gave rise to two new sets of problems. First of all, the presence of three new coastal states in the area implied that a good number of new delimitation agreements needed to be concluded in areas were none had existed before. Secondly, the question arose as to the legal validity of maritime boundary agreements concluded in the past by states, which were no longer in control of the area concerned.

2.3.1   New Agreements

As the former Soviet Union had not delimited the maritime spaces between its former republics because their legal regime remained a Union competence, new boundaries needed to be created. When moving from the south to the north, it concerned the boundaries between the Russian Federation (Kaliningrad region) and Lithuania, Lithuania and Latvia, Latvia and Estonia, and finally Estonia and the Russian Federation respectively. If in the summer of 1991 one could state that a maritime boundary had been agreed upon in the whole Baltic Sea area with the exception of the area south and southeast of the Danish island of Bornholm and most of the tripoint areas,[45] this certainly was no longer a true statement after 1991 had come to a close.

Most of these agreements have been concluded in the meantime: In chronological order it concerns Estonia-Latvia in 1996,[46] Lithuania-Russian Federation in 1997,[47] Latvia-Lithuania in 1999,[48] and Estonia-Russian Federation in 2005.[49] The

---

[45] E. Franckx "Maritime Boundaries and Regional Cooperation in the Baltic" (1992) 20 (issue 1–3) *International Journal of Legal Information* 18.

[46] Agreement on the Maritime Delimitation in the Gulf of Riga, the Strait of Irbe and the Baltic Sea of 12 July 1996 (2001) 1955 UNTS 345, 355–360.

[47] Two treaties were concluded on the same day: 1) Treaty on the Lithuanian-Russian State Border of 24 October 1997, reprinted in an unofficial English translation by the present author in E. Franckx "New Maritime Boundaries Concluded in the Eastern Baltic Sea Since 1998" (2001) 16 (issue 4) *International Journal of Marine and Coastal Law* 645, 655–655; and 2) Treaty on the Delimitation of the Exclusive Economic Zone and the Continental Shelf in the Baltic Sea of 24 October 1997, reprinted in English translation in E. Franckx "Two More Maritime Boundary Agreements Concluded in the Eastern Baltic Sea in 1997" (1998) 13 (issue 2) *International Journal of Marine and Coastal Law* 274, 282–283.

[48] Agreement on the Delimitation of the Territorial Sea, Exclusive Economic Zone and Continental Shelf in the Baltic Sea of 9 July 1999, reprinted in English translation in Franckx, New Maritime Boundaries, note 47 at 657–658.

[49] Treaty on the Delimitation of the Maritime Areas in the Gulf of Narva and the Gulf of Finland of 18 May 2005, reprinted in an unofficial French translation by the present author in E. Franckx and

latter stands out, as the Russian Federation withdrew its signature on 6 September 2005, after an introductory declaration had been added by the Estonian Parliament to the law of ratification in which the legal continuity of the Estonian Republic proclaimed in 1918 and a special reference to the Tartu Peace Treaty and its boundary provisions were mentioned, two issues carefully avoided in the text of the delimitation agreement itself.[50] The two states again signed the slightly amended document a second time in 2014,[51] but ratification is still pending. All the other just-mentioned agreements have also entered into force, except for the one between Latvia and Lithuania.

2.3.2 Existing Agreements

A first set of issues relating to the fate of previously concluded maritime boundary agreements in the Baltic Sea arose with respect to the reunification of Germany. It certainly made one boundary agreement, concluded in the past between its two subentities *inter se*, simply redundant.[52] At the same time, the recognition of the maritime boundary concluded by the former German Democratic Republic with Poland[53] caused some concern because of the disputed land frontier on which it was based. But as was the case with all other maritime boundaries concluded by the former German Democratic Republic in the past, also the maritime boundary with Poland was formally recognized by the newly unified German state.

The situation with respect to Latvia, Lithuania and Estonia was of a totally different nature because the countries claimed that their legal status, as far as maritime boundaries were concerned, should return to the situation *ante quo*, meaning before the illegal annexation by the former Soviet Union. They considered themselves not bound by the maritime delimitation agreements concluded in the past by this country in areas where they now had acquired sovereignty or sovereign rights according to the UNCLOS.[54] Finland and Sweden, located on the opposite side, were rather of the opinion that these maritime boundary agreements concluded in the past by the former Soviet Union still governed the maritime delimitation in question. The way out of this dilemma for the countries involved was that new agreements were

---

M. Kamga "L'existence éphémère du Traité de délimitation maritime entre la République d'Estonie et la Fédération de Russie en mer Baltique" (The Short Existence of the Maritime Delimitation Agreement in the Baltic Sea Between the Republic of Estonia and the Russian Federation) (2008) 12 *Annuaire du Droit de la Mer 2007* 393, 421–423.

[50] On this highly exceptional practice of withdrawing one's signature as applied to this particular delimitation agreement, see ibid., 394–395 and 404–407.

[51] Information available at <https://www.dur.ac.uk/ibru/news/boundary_news/?itemno=20200&re href=%2Fibru%2Fnews%2F&resubj=Boundary+news+Headlines>.

[52] It concerns the 1974 Protocol (note 36).

[53] 1968 Treaty (note 35).

[54] See for instance the 1988 Agreement (note 39).

concluded. These agreements, while not containing any explicit reference to the previously concluded agreements, nevertheless arrive at exactly the same delimitation line. The most recent of such agreements is the one concluded between Lithuania and Sweden in 2014.[55]

The overview of this fourth period can be wrapped up by making reference to the conclusion of two more tripoint agreements. The first one concerns the tripoint between Estonia, Latvia and Sweden,[56] the second one closed the remaining gap between Estonia, Finland and Sweden.[57] A third one between Lithuania, the Russian Federation and Sweden seems to be in the making.

2.4   Remaining Gaps

After having summarized what has been achieved so far, it is now time to shift the attention to what still remains to be done with respect to the delimitation of maritime spaces in the Baltic Sea.

Two presumptions will guide the analysis of the present part. The first is that states will continue to settle the delimitation of maritime areas between them either on a bilateral basis, when the delimitation concerns two states, or on a trilateral basis if tripoint areas are involved. If the past can be relied upon to predict the future, the likelihood that any third party settlement will be relied upon remains rather slim. The second presumption is that states will only start negotiations on the remaining tripoints when the bilateral negotiation process between them has been concluded with the signing of the relevant bilateral agreements first.

One major segment requiring bilateral action stands out in this respect, namely the area south and southeast of the Danish island of Bornholm. Denmark and Poland have not been able to sign any document relating to this area.

All the other areas requiring bilateral action have been covered, be it that between Lithuania and Latvia a bilateral document has been signed but is still awaiting ratification more than 15 years after signature. A somewhat similar situation exists between Estonia and the Russian Federation, but here, for reasons explained above,[58] two signatures intervened and depending on which one is used as a starting point for the calculation, the period between signature and ratification amounts to either 11 or 3 years.

---

[55] Agreement on the Delimitation of the Exclusive Economic Zones and the Continental Shelf in the Baltic Sea of 10 April 2014, not yet published in UNTS but English translation already available   at   <https://treaties.un.org/doc/Publication/UNTS/No%20Volume/53412/Part/I-53412-080000028045a3bc.pdf>.

[56] Agreement on the Common Maritime Boundary Point in the Baltic Sea of 30 April 1997, reprinted in English translation in Franckx, Two More Boundary Agreements, note 47 at 281.

[57] Agreement on the Common Maritime Boundary Point in the Baltic Sea of 16 January 2001 (2011) 2474 UNTS 43, 45–46.

[58] See note 50 and accompanying text.

As far as the trilateral agreements are concerned, three different situations can be distinguished, namely areas where one of the required bilateral agreements is missing, areas where all bilateral agreements have been signed, but one of them is missing ratification, and finally areas where all bilateral treaties have been signed and ratified.

2.4.1   One of the Required Bilateral Agreements Is Missing

As previously explained, there exists only one area in the Baltic Sea where one of the required bilateral agreements is missing, namely between Denmark (island of Bornholm) and Poland. It involves the tripoints between Denmark, Germany and Poland as well as the tripoint between Denmark, Poland and Sweden. According to the second presumption, it is believed that these tripoints will not be tackled in any foreseeable future as long as no further progress is made on the bilateral level.

2.4.1.1   One of the Required Bilateral Agreements Has Been Signed, But Not Ratified

Under this sub-heading two agreements need to be mentioned. First, the one concluded between Latvia and Lithuania in 1999.[59] The formal adoption of the tripoint between Latvia, Lithuania and Sweden is consequently put on hold. A similar situation exists as a result of the non-ratification of the maritime boundary agreement between Estonia and the Russian Federation,[60] implying that also the tripoint between Estonia, Finland and the Russian Federation in the Gulf of Finland remains to be settled at present. It is submitted that one will have to await ratification of these two agreements before the relevant tripoints will be fixed by agreement.

2.4.1.2   All Required Bilateral Agreements Have Been Duly Signed and Ratified

This last category considered here concerns those areas where all the bilateral delimitation lines are operational, but where the exact location of the remaining tripoints still needs to be determined by agreement. It concerns two tripoints in the southwestern Baltic Sea, namely a first one between Denmark, Germany and Sweden and a second one between Denmark (island of Bornholm), Germany and Sweden, as well as the tripoint in the southern Baltic Sea between Lithuania, the

---

[59] 1999 Agreement (note 48).

[60] 2005 Treaty (note 49), as slightly modified in 2014 (note 51 and accompanying text).

Russian Federation and Sweden. Of the three categories here distinguished, this one seems to be the most promising for observing concrete developments in the near future.[61]

2.5   Conclusions

If one compares the situation as it existed just before the dissolution of the former Soviet Union[62] and today, the same general conclusion seems to be justified, namely that the Baltic Sea remains a regional sea where most of the maritime boundaries have been delimited. Still the same area is not covered by any bilateral agreement, namely south and southeast of the Danish island of Bornholm. This area forms part of the Danish-Polish trough, which connects the potentially petroleum-rich southern part of the Baltic with the definitively petroleum-rich North Sea.[63] Also the fact that this island received full effect when Denmark delimited its northern maritime boundary with Sweden in 1984, but only limited effect when its southwestern boundary was settled with the former German Democratic Republic in 1988,[64] can be subsumed as having contributed to this delay. For the rest, only restricted areas surrounding tripoints remain to be settled between the parties. The latter number seven in total. This is slightly higher than the situation as it existed at the end of 1991, but given the many new tripoints that were created after the emergence of three new coastal states, this number must certainly be placed in perspective.

The general conclusion therefore seems to be justified that the *status quo ante* has been restored, and that the Baltic Sea has regained a degree of completeness as far as agreed maritime boundaries is concerned, an achievement that is not easily found in other regions of the world.

Consequently, the research question formulated in the beginning of this article, namely whether the general submission that the Baltic Sea is probably the most regulated international marine area also applies to maritime delimitation, can be answered positively. Whether all the remaining gaps will be filled in the near future, remains somewhat uncertain, especially given the general deterioration of the relationship between the European Union and the Russian Federation in recent times.

---

[61] As already alluded to above with respect to the tripoint negotiations being held between Lithuania, the Russian Federation and Sweden.

[62] Franckx, note 45 at 18.

[63] E. Franckx "Denmark-German Democratic Republic (Report Number 10–11)" in J.I. Charney and L.M. Alexander (eds) *International Maritime Boundaries*, Vol. 2, (Martinus Nijhoff, Dordrecht, 1993) 2087, 2089.

[64] E. Franckx "Region X: Baltic Sea Maritime Boundaries" in J.I. Charney and L.M. Alexander (eds) *International Maritime Boundaries*, Vol. 1, (Martinus Nijhoff, Dordrecht, 1993) 345, 359.

Chapter 3
Straits in the Baltic Sea: What Passage Rights Apply?

Pirjo Kleemola-Juntunen

Abstract This chapter studies the navigational rights in the two major Baltic Sea straits, the Åland Strait and the Danish Straits, in the light of existing international law. Both straits are of major commercial and strategic relevance in the Baltic Sea and both have been subject to specific treaties in the past. Yet the current legal status of both straits include several uncertainties with regard to the extent to which other states' ships may freely navigate therein. The examples hence illustrate that even precise regulation in the law of the sea and specific international conventions is not a guarantee for uniformity of interpretation and application of navigational rights in the Baltic Sea.

Keywords International straits · Transit passage · Danish straits · Strait of Åland · Navigational rights

3.1 Introduction

The purpose of this paper is to study navigational rights in the major Baltic Sea straits, the Åland Strait and the Danish Straits, in the light of the provisions of the 1982 UN Convention on the Law of the Sea (UNCLOS). The Baltic Sea is a semi-enclosed sea surrounded by nine countries. A number of important international navigation routes pass through its waters. The navies of both bordering and non-bordering states cruise the Baltic Sea, creating an area of delicate political balance

---

Jur.Dr., Postdoc Researcher. This chapter is written as a part research project "Demilitarisation in an increasingly militarised world. International perspectives in a multilevel framework – the case of the Åland Islands". The research project is a co-operation between the Åland Islands Peace Institute (ÅIPI) and The University of Lapland and its Arctic Center in Rovaniemi (Finland). The Project is funded by the KONE Foundation.

P. Kleemola-Juntunen (✉)
Arctic Centre/NIEM, University of Lapland, Rovaniemi, Finland
e-mail: pirjo.kleemola-juntunen@ulapland.fi

© Springer International Publishing AG, part of Springer Nature 2018
H. Ringbom (ed.), *Regulatory Gaps in Baltic Sea Governance*, MARE
Publication Series 18, https://doi.org/10.1007/978-3-319-75070-5_3

21

and major strategic importance. There are two straits of major commercial and strategic relevance in the Baltic Sea namely the Åland Strait and the Danish Straits (which includes the Sound, the Great Belt and the Little Belt).[1]

Free and open access to the ports of the Baltic Sea for purposes of commerce, national defense and for other forms of international cooperation have been and remain in the interests of third states. Therefore, it is important to balance the interests of coastal states and third state ocean users. The will of powerful states has had a significant effect on the development of passage through the straits. States have promoted their interests by means of treaty arrangements, such as the 1857 Treaty on the Sound Dues with Denmark, or through the use of force.[2] The 1921 Geneva Convention relating to the Non-fortification and Neutralisation of the Åland Islands that covers parts of the Åland Strait belonging to Finland was concluded between interested states, with the intention of reducing the islands' potential to be used for military purposes. The straits constitute part of the sea and they are important for international navigation because ships need straits for navigating between home ports and oceans. Notwithstanding the existence of the provisions of international law concerning various zones within the ocean, the straits are governed by specific regulations as a result of different conventions on the law of the sea. However, certain regulations are only applied to straits that are international. If a strait is completely situated within the internal waters of a given state, then it is not "international".

With respect to the legal requirement of a strait in UNCLOS, a strait must be used for international navigation between one part of the high seas or an EEZ and another part of the high seas or an EEZ. Do the Åland Strait and the Danish Straits fulfil the requirements necessary to be defined as a strait used for international navigation? Geographically, the Åland Strait is situated between Sweden and the archipelagic region of the Åland Islands; the Sound is situated between Sweden and Denmark. They are both narrow sea areas connecting two areas of the seas. With respect to the functional aspect of a strait, however, it must be emphasized that traffic through the Åland Strait and the Danish Straits is so large that they can both be considered to be straits used for international navigation. Hence, the Åland Strait and the Danish Straits meet the legal requirements of a strait.

---

[1] According to A.G. López Martín *International Straits, Concept, classification and Rules of Passage* (Springer, 2010) 203 the Entrance to the Gulf of Finland that connects the Baltic Sea and the Gulf of Finland and the territorial sea of Russia is also a strait in the Baltic Sea. López Martín considers the strait to be the objective of art. 45(1)(b).

[2] One example is the use of force by Japan in 1863, when Japan opposed passage through the Shimonoseki Strait. Another is Turkey in 1774, when the prohibition of passage was mitigated in favour of merchant vessels in the peace treaty between Russia and Turkey. E. Brüel *International Straits, A Treatise on International Law, Vol. I., The General Legal Position of International Straits* (Sweet & Maxwell, London, 1947) 102–105; G. Alexandersson *The Baltic Straits* (Marinus Nijhoff Publishers, 1982) 72–73; C.L. Rozakis and P.N. Stagos *The Turkish Straits* (Martinus Nijhoff Publishers, 1987) 82.

The conventions relating to the Danish Straits and the Åland Strait are old and therefore this chapter will first look at the time prior to UNCLOS, the time when only one concept, the right of innocent passage, applied to the passage through the straits. The principle of innocent passage developed into customary international law and the straits were considered to be part of the territorial sea until the 1958 Convention on the Territorial Sea and Contiguous Zone established a different legal status for international straits. This chapter first gives an overview of the legal regime governing straits used for international navigation prior to UNCLOS in section two and then section three summarizes how the passage rights are regulated under UNCLOS. Even though the chapter aims to examine currently valid rules, it is important to understand the historical process pertaining to such rules. Section four of this chapter then outlines the differences between transit passage and innocent passage. In addition to the concept of transit passage, UNCLOS created another concept related to the passage through the straits, namely "straits governed by "long-standing conventions in force"". Although the article does not mention the straits, on the basis of the UNCLOS negotiations these straits are taken to be the Turkish Straits, Danish Straits, the Strait of Magellan and the Åland Strait. Each of these exceptional straits have *sui generis* regimes governed by "long-standing conventions in force". Two of these four exceptional straits are located in the Baltic Sea, the Åland Strait and the Danish Straits. Section five examines the Åland Strait and the Danish Straits as straits to which art. 35(c) of UNCLOS applies. Even though other states did not question their position during the Conference, it still remains open to interpretation on some issues. This is important since third states may enter the Baltic Sea and are entitled to use it in accordance with the law of the sea. Interpretative positions are examined in Section six. Section seven draws conclusions.

3.2 The Law Prior to UNCLOS

Before the 1982 UN Convention on the Law of the Sea, there was only one concept regulating the passage through territorial sea and straits, innocent passage. During the end of the twentieth and the beginning of the twenty-first centuries learned societies and researchers made proposals to codify the right of innocent passage. All these attempts failed.

Regarding the current legal regime on straits, the first important ruling was on the *Corfu Channel*. In its judgment of 9 April 1949 on the *Corfu Channel* case the International Court of Justice held that it was in accordance with international custom that in time of peace warships had the right to innocent passage through international straits if not otherwise prescribed in an international convention.[3] In the 1947 memorial the United Kingdom stated that warships had the right of innocent passage through territorial seas, which was an *a fortiori* right when straits were

---

[3] *Corfu Channel* case, Judgment of April 9th, 1949: ICJ Reports 1949, 4, 28.

concerned. The United Kingdom made a reference to previously published regulations by a large majority of states, as well as an inventory of state practices leading up to the 1930 Hague Conference. By doing so the United Kingdom managed to portray the extent of how the right of innocent passage through territorial waters had been acknowledged by a majority of states.[4] Furthermore, the United Kingdom stated that the majority of states that had replied to a questionnaire on the issue had expressed their views and had agreed that warships have a right of innocent passage through the territorial waters of another state. Furthermore, it is noteworthy that the United Kingdom, as party to 1921 Åland Convention, stated that the right of innocent passage through territorial waters for warships had been given former recognition in art. 5 of the 1921 Åland Convention, which explicitly reserved the right.[5] It is apparent that the *Corfu Channel* case had created a different legal status for international straits.

Eventually, as a result of the First Law of the Sea Conference in 1958 the Convention on the Territorial Sea and Contiguous Zone established the right of innocent passage through the territorial sea of a coastal state for the ships of foreign states. Also, on straits the Convention said that "there shall be no suspension of the innocent passage of foreign ships through straits which are used for international navigation between one part of the high seas and another part of the high seas or the territorial sea of a foreign state".[6] Navigation through international straits was considered to be so important that special rules were necessary. However, one must also bear in mind that while coastal states were prohibited from suspending innocent passage through straits used for international navigation, the Convention also left it up to states to decide whether the nature of passage was innocent or not.[7] The Convention does not mention warships but on the ground of the location of provisions in the subsection covering all ships and the *Corfu Channel* case it was interpreted that warships enjoy the right of innocent passage through international straits.

---

[4] See arts. 4, 12 of the Draft Report, Circulated to the Members of the Committee on 3 April 1930, (Work of the First Sub-committee) in S. Rosenne (ed) *League of Nations, Conference for Codification of International Law (1930)* (Oceana Publications, 1975) 1405, 1408.

[5] *Corfu Channel* case, Section B. Written Statements, The North Corfu Channel, 1. Memorial Submitted by the Government of the United Kingdom of Great Britain and Northern Ireland, ICJ Reports 1949, 43.

[6] Art. 14(1), 16(4).

[7] Art. 14(4) says that passage is innocent so long as it is not prejudicial to the peace, good order or security of the coastal state. Such passage shall take place in conformity with these articles and with other rules of international law.

3.3   The 1982 UNCLOS Regime of Passage Through Straits Used for International Navigation

During the Third Law the Sea Conference in 1974–1982 many of the problems related to innocent passage and freedom of navigation tended to focus on the regime of straits. Major naval powers, especially the then superpowers, were primarily concerned about the potential effects of introducing 12 nautical mile wide territorial seas and the abolishment of certain high sea corridors through straits with widths that were less than 24 nautical miles. As a result of tensions over these issues, a confrontation between coastal states and naval powers started to emerge[8]; one side requiring a similar right of passage through straits as on the high seas and the other side requiring that the straits had a status similar that of a territorial sea.[9] To balance the competing interests of the international community, the UNCLOS created a concept, "transit passage", which is applicable in straits which are used for international navigation. States with substantial navies wanted to retain passage rights for their vessels to the fullest extent possible in spite of new maritime zones being created by UNCLOS. Furthermore, the purpose of "transit passage" was to ensure that free transit through international straits was provided and to restrain states bordering straits from interfering with transit. Previously the 1958 Territorial Sea Convention has provided foreign vessels with the right to non-suspendable innocent passage, which included some limitations on innocent passage.[10]

If a state controls transit through a strait it means it can hinder or prohibit traffic completely. In closing straits, states can prevent offensive movements approaching from beyond the strait. Closing straits can also be useful when attempting to hinder a foreign vessel's access to a coastal state's territorial seas.[11] Therefore, it is easy to understand why the possibility of wider territorial seas was a concern for major naval powers, as well as other seafaring nations. The preservation of free passage

---

[8] H. Caminos "The Legal Régime of Straits in the 1982 United Nations Convention on the Law of the Sea" 205 *Recueil Des Cours/ Hague Academy Collected Courses* 1987 V (Martinus Nijhoff Publishers, 1989) 13–245 at 63; López Martín, note 1 at 29–30; Kennedy, "A Brief Geographical and Hydrographical Study of Straits which Constitute Routes for International Traffic," UN Doc. A/CONF.13/6 and Add.1 (1957) in UNCLOS I, Official Records, Volume I, 114–164; M.F. Maduro "Passage through International Straits: The Prospects Emerging from the Third United Nations Conference on the Law of the Sea" (1980) 12 *Journal of Maritime Law and Commerce* 65–95 at 65–69. It was estimated that widening the territorial sea to 12 nautical miles would enclose some 116 international straits within territorial waters, e.g. Gibraltar, Bab al-Mandeb, Malacca.

[9] C. A. Fleischer "International Straits: A Key Issue at the Law of the Sea Conference" (1975) 1 *Environmental Policy and Law* 120–126 at 122. UN Doc. A/AC/138/SC II/L.4 81,971), UN.Doc. A/CONF.62/C.2/L.11 (1974) the 1974 Caracas Session (Bulgaria and others e.g. USSR).

[10] Overflight is prohibited above the straits and submarines must navigate on the surface in straits, which is the procedure for passage through territorial seas. H. Caminos and V. P. Cogliati-Bantz *The Legal Regime of Straits* (Cambridge University Press, 2014) 38.

[11] Brüel, note 2 at 25.

through straits was so crucial to some states that the success of the Convention very much depended on this issue being resolved.[12]

3.3.1 Transit Passage

A regime of transit passage for straits, including the delimitation of the sea area to which it applies is established and defined in Section 2, arts. 37–44. These articles cover the rights and duties of states bordering straits and states using the right of transit passage. Section 3 only comprises one article (art. 45), which covers non-suspendable innocent passage and specifies the conditions for when innocent passage in the straits can be applied to international navigation.

The scope of the regime of transit passage is limited by art. 37 "to straits used for international navigation between one part of the high seas or an exclusive economic zone and another part of the high seas or exclusive economic zone". The term "straits used for international navigation" as a definition is still ambiguous because the Third Law of the Sea Conference could not agree on the meaning of the term. The geographical and functional concepts related to the term itself could be instrumental in defining its meaning. However, even these concepts are ambiguous because no new legal agreement has been made to clarify whether one concept predominates the other.[13] The *Corfu Channel* case can offer some form of guidance on the matter, although the case refers to "a useful route for international navigation", which is a more liberal term of expression.[14]

The right of transit passage is defined in art. 38. According to this article all ships and aircraft enjoy a right of transit passage, which cannot be impeded. The article provides that transit passage is only exercised for the purpose of continuous and expeditious transit. The only exception to the rule of continuous and expeditious transit is when "entering, leaving or returning from a State bordering a strait, subject to the conditions of entry to that State".[15] As a result of the application of the new concept of transit passage, the legal regime of many straits would have changed because it was accepted that this new regime would also cover straits where the right of innocent passage had previously applied. This means that the regime of innocent passage is exceptional, and applicable only within a limited number of straits.

---

[12] E.J. Manner "Some Observations on the Effects and Applications of the New Law of the Sea, with Special Reference to the Baltic" in M. Tupamäki (ed) *Essays on International Law* (Finnish Branch of the International Law Association, Vammala, 1987) 114–144 at 120.

[13] *Corfu Channel* case, ICJ Reports 1949 at 28, Caminos, note 8 at 142–143.

[14] *Corfu Channel* case, ICJ Reports 1949 at 28–29.

[15] Caminos, note 8 at 143–144, UNCLOS art. 38(2).

3.3.2 Exceptions

Although the regime of transit passage is the main rule applied in straits used for international navigation the Convention mentions several exceptions to the transit passage.[16] Firstly transit passage is not applied in a strait where high sea routes or EEZ routes run through the middle of the strait nor internal waters within a strait.[17] However, when the establishment of strait baselines has an effect of enclosing internal waters areas that were never previously considered as such, transit passage rights are preserved. Secondly art. 35 (c) creates a special status for straits that are governed "in whole or in part" by long-standing international conventions in force. According to art. 35 (c), provisions on transit passage do not affect the legal regime of straits regulated by conventions. The aim of this exception is to preserve the regimes established by the conventions relating to the Åland Strait, the Danish Straits, the Turkish Straits, and the Strait of Magellan.[18]

The third exception is a strait where non-suspendable innocent passage continues. Art. 38 (1) of UNCLOS refers to a strait, which is formed by an island (belonging to a state situated adjacently to the strait in question) and its mainland. An example of this type of strait is the Messina Strait, which is located between Italy and Sicily.[19] Non-suspendable innocent passage continues through straits which are situated between a part of the high seas or EEZs and a territorial sea of a foreign state. This category is regulated by art. 45 (1) (b). An example of this type would be the Strait of Juan de Fuca, which is located between the United States and Canada and the Straits of Tiran between the Red Sea and the Gulf of Aqaba.[20]

3.3.3 Innocent Passage vs. Transit Passage

Hypothetically speaking, the regime relating to transit passage could have fundamentally altered navigation through the Åland Strait and the Danish Straits, especially where warships and military aircraft were concerned. Had this happened, submarines would have been allowed to navigate, submerged, through the strait,

---

[16] H. Caminos and V. P. Cogliati-Bantz, note 10 at 42; UNCLOS arts. 35, 36, 38 and 45.

[17] Art. 35 (a), 36.

[18] UNCLOS III, Official Records, Volume XIV, 21, 35; K. Hakapää *Marine Pollution in International Law, Material Obligations and Jurisdiction with special Reference to the Third United Nations Conference on the Law of the Sea* (Suomalainen Tiedeakatemia, Helsinki, 1981) 202; H. Caminos and V. P. Cogliati-Bantz, note 10 at 71.

[19] S. Mahmoudi "Transit Passage" *Max Planck Encyclopedia of Public International Law* available at MPEPIL online/Transit Passage <http://www.mpepil.com> para. 16.

[20] Ibid.; K. Buntoro *An Analysis of Legal Issues relating to Navigational Rights and Freedoms through and over Indonesian Waters* (Australian National Centre for Ocean Resources and Security (ANCORS) - Faculty of Law, University of Wollongong 2010) available at <http://ro.uow.edu.au/theses/3091>.

military aircraft would have had the right of overflight and aircraft carriers could have exercised their activities without restrictions. All vessels and aircraft would have enjoyed passage through the Åland Strait and the Danish Straits in a large portion of the sea that would have extended from the North Sea across the Baltic Sea to the most northern part of the Gulf of Bothnia. As a result, the regime of transit passage would have altered the ability of states bordering these narrow straits to protect themselves against maritime safety and security threats because it reduces their ability to regulate passage.

The 1958 Convention on the Territorial Sea and Contiguous Zone regulated passage through territorial seas and straits used for international navigation by providing for the right of innocent passage. However, as regards the straits there was a special regulation, a coastal state was not allowed to suspend innocent passage through the strait. During the preparatory sessions of the First Law of the Sea Conference in 1955, the parties discussed the fact that a number of international straits were covered by special conventions and each of those conventions provided for a different regime.[21] Paragraph five of the commentary on art. 24 covering passage of warships says: "The article does not affect the rights of States under a convention governing passage through the straits to which it refers." It seems clear that it was not the intention of the Commission to interfere in any way with the special conventions already covering passage in straits. However, this general principle did not end up in the Convention itself.[22] The Convention on the Territorial Sea and Contiguous Zone created a separate legal status for the straits used for international navigation which was further elaborated on by UNCLOS and also in terms of straits regulated by international conventions.

As a result of the extension of territorial seas to 12 nautical miles, straits used for international navigation that were previously subject to the high seas freedom of navigation may fall within the territorial seas of one or more coastal states.[23] At the Third Law of the Sea Conference navy states were particularly concerned about changes to their navigational rights as a result of the extension of territorial seas because they wanted to retain their navigational freedom to the highest possible degree. Thus, there was a need for a new legal regime that, unlike innocent passage, regulated the passage of all ships and the overflight of aircraft in the straits. .[24]

The scope of art. 38 covers all ships and aircraft, whereas no distinction between their nationality and ownership or legal status (merchant ship or warship, civil or state aircraft) is provided. The right of transit passage is applied to all types of vessels and aircraft, regardless of their legal status. Moreover, the wording of the article effectively excludes the possibility of a coastal state asking for prior-authorization or

---

[21] *Yearbook of the International Law Commission*, Summary records of the seventh session 2 May — 8 July 1955, 1955, Volume I, 150.

[22] *Yearbook of the International Law Commission* Documents of the eighth session including the report of the Commission to the General Assembly, 1956, Volume II, 277.

[23] M. Reisman "The Regime of Straits and National Security: an Appraisal of International Law-making" (1980) 74 *American Journal of International Law* 48–76 at 30, 67.

[24] Caminos, note 8 at 122; UNCLOS art. 38(2).

prior-notification when vessels undertake passage through their territorial seas.[25] Another important term used is "freedom of navigation and overflight," which links passage through straits to navigation in the high seas. The expression is also found in the Geneva Convention on the High Seas.[26] Arts. 38 and 39 create separate rights and obligations. Any violation of duties of vessels and aircraft included in art. 39 does not mean that a vessel forfeits its right to transit passage. This viewpoint is supported in arts. 42 (2) and 44, which state that bordering straits shall not impede or interfere with the right of transit passage in any way whatsoever.[27] Matters related to state security also highlight a difference between transit passage and innocent passage. States bordering straits cannot suspend transit passage because of security concerns, whereas innocent passage can be considered non-innocent if it is prejudicial to the security interests of coastal states.[28] Nevertheless, the right of transit passage may not exclude the right of self-defence.[29] Overall, it seems there is quite an obvious difference regarding the jurisdictions of the different states that border straits, depending on the type of passage rights that are used, i.e. transit or innocent passage.

Although the extent of state jurisdiction within straits is hard to measure, UNCLOS grants states bordering straits the right to enforce laws and regulations in order to govern transit passage. States bordering straits may regulate traffic in straits by prescribing sea lanes and traffic separation schemes but they need to be adopted by the IMO. Also, the laws and regulations relating to the pollution of straits must comply with generally accepted regulations.[30]

Furthermore, although states must refrain from enforcing laws and regulations that might cause a disturbance to transit passage through straits, the Convention does not give any information or guidelines on how to actually do so. Overall, it seems that art. 44 restricts states from hampering or suspending transit passage through their straits. When comparing the jurisdiction of states bordering straits on matters concerning transit passage with matters concerning innocent passage, it would appear that state competence in relation to transit passage is significantly limited.[31]

---

[25] Caminos, note 8 at 144.

[26] Art. 2, Convention on the High Seas, 450 UNTS 11; Finnish Treaty Series (SopS 6–7/1965).

[27] Caminos, note 8 at 150.

[28] UNCLOS arts. 19 (2) and 44; R.R. Churchill and A.V. Lowe *The Law of the Sea* (Manchester University Press, Manchester, 1999) 107.

[29] Military and Paramilitary Activities in and against Nicaragua, ICJ Reports 1986 at 14, 102–104, 110–111; Churchill and Lowe, note 28 at 107, see also J. Hargrove "The Nicaragua Judgment and the Future of the Law of Force and Self-Defense" (1987) 81 *American Journal of International Law* 135–143.

[30] UNCLOS art. 42 (1).

[31] M.H. Nordquist *et al.* (eds) *United Nations Convention on the Law of the Sea 1982, A Commentary, Volume II* (Martinus Nijhoff Publishers, 1993) 375, 388; Churchill and Lowe, note 28 at 108; S.N. Nandan "Legal Regime for Straits Used for International Navigation" in B. Öztürk and R. Özkan (eds) *The Proceedings of the Symposium on the Straits Used for International Navigation, 16–17 November 2002, Ataköy Marina, Istanbul – Turkey, Publication Number: 11* (Turkish Marine Research Foundation, Istanbul, 2002) 1–11 at 6; D.R. Rothwell and T. Stephens *The International Law of the Sea* (Hart Publishing Ltd., 2010) 243.

Therefore, the concept of transit passage changes the legal regime of straits when passage is concerned. States had previously agreed that all states enjoyed freedom of passage through waters considered to be international straits. Nevertheless, if a strait was twice as narrow as the breadth of a territorial sea then passage was subject to the regime of innocent passage, which meant that passage was non-suspendable.[32] In straits like these coastal states are given the discretion to consider whether passage is non-innocent, which empowers them with the right to suspend passage. A similar provision on innocent passage is included in UNCLOS, which provides more detail about what kinds of activities are considered prejudicial to the peace, good order or security of the coastal state.[33] With the widening of territorial seas came the expansion of coastal state jurisdiction, too, but coastal states' jurisdiction was encroached on in relation to straits used for international navigation because, in these types of straits, the right of transit passage is applied instead of the right of innocent passage. The change within the legal regime of straits used for international navigation is significant because the concept of transit passage is applied to straits that are too narrow to include a corridor through the EEZ or the high seas. It is easy to relate this change to another similar development in the regime of transit

---

[32] Art. 14 (4) of the 1958 TSC prescribes that passage is innocent so long as it is not prejudicial to the peace, good order or security of the coastal state and art. 16(4) states that: "There shall be no suspension of the innocent passage of foreign ships through straits which are used for international navigation between one part of the high seas and another part of the high seas or the territorial sea of a foreign State." R.B. McNees "Freedom of Transit through International Straits" (1974) 6 *Journal of Maritime Law and Commerce* 175–211 at 185–186.

[33] Art. 19 of UNCLOS states that

"1. Passage is innocent so long as it is not prejudicial to the peace, good order or security of the coastal State. Such passage shall take place in conformity with this Convention and with other rules of international law.

2. Passage of a foreign ship shall be considered to be prejudicial to the peace, good order or security of the coastal State if in the territorial sea it engages in any of the following activities:

(a) any threat or use of force against the sovereignty, territorial integrity or political independence of the coastal State, or in any other manner in violation of the principles of international law embodied in the Charter of the United Nations;

(b) any exercise or practice with weapons of any kind;

(c) any act aimed at collecting information to the prejudice of the defence or security of the coastal State;

(d) any act of propaganda aimed at affecting the defence or security of the coastal State;

(e) the launching, landing or taking on board of any aircraft;

(f) the launching, landing or taking on board of any military device;

(g) the loading or unloading of any commodity, currency or person contrary to the customs, fiscal, immigration or sanitary laws and regulations of the coastal State;

(h) any act of wilful and serious pollution contrary to this Convention;

(i) any fishing activities;

(j) the carrying out of research or survey activities;

(k) any act aimed at interfering with any systems of communication or any other facilities or installations of the coastal State;

(l) any other activity not having a direct bearing on passage."

Art. 45(2) states that: "There shall be no suspension of innocent passage through such straits."

rights, namely that the regime has been expanded to include the right of overflight above straits.

Selecting the high seas term was clearly intentional in order to emphasize the difference between the two regimes of transit passage and innocent passage through territorial seas and the sea areas of archipelagic states. Articles concerning innocent passage through territorial seas or archipelagic waters do not include any mention of freedom of navigation and overflight.[34] It is also noteworthy that art. 34 mentions that passage through straits used for international navigation shall not affect the legal status of waters forming straits. Furthermore, states bordering straits will not be affected by art. 34 as regards their sovereign jurisdiction over air space, sea bed and subsoil areas making up their waters. However, art. 34's second paragraph establishes the legal limits of a coastal state's sovereign jurisdiction if they share their borders with international straits. The provisions found in this part of art. 34 prescribe that state sovereignty and jurisdiction is regulated by the provisions of straits used for international navigation and other rules of international law. Therefore, any activities that are unrelated to passage remain subject to the other applicable provisions of the Convention. Provisions on the right of transit passage are not applicable to the internal waters of coastal states that are connected to straits. However, an exception to this rule is made for sea areas that had never previously been considered internal waters.[35]

During the Third Law of the Sea Conference Denmark, Finland and Sweden each suggested that it would be within their mutual interests to regulate navigation through the straits according to a legal norm that excludes provisions on transit passage. There was no doubt whether the Danish Straits and the Åland Strait would qualify as "straits used for international navigation", the term officially used in draft articles and the eventual 1982 Convention. However, this conclusion raised problems for states bordering the Danish Straits and the Åland Strait.[36] Art. 35 (c) of UNCLOS was the result of such a compromise, according to which nothing affects "the legal regime in straits in which passage is regulated in whole or in part by long-standing international conventions in force specifically relating to such straits".[37] However, the straits that are governed by art. 35 (c) are not listed in the article exclusively. While signing UNCLOS, Finland issued the following Declaration: "It is the understanding of the Government of Finland that the exception from the transit passage regime in straits provided for in art. 35 (c) of the Convention is applicable to the strait between Finland (the Åland Islands) and Sweden. Since in that strait the passage is regulated in part by a long-standing international convention in force, the present legal regime in that strait will remain unchanged after the entry into force of the Convention." Sweden issued a corresponding Declaration. Finland and Sweden

---

[34] J. A. de Yturriaga *Straits Used for International Navigation, A Spanish Perspective* (Martinus Nijhoff Publishers, 1991) 167.

[35] Manner, note 12 at 121; Nordquist *et al.* (eds), note 31 at 298.

[36] Manner, note 12 at 123.

[37] UNCLOS III, Official Records, Volume XIV, 21, 35, 61, UN Doc. A/CONF.62/SR.135, UN Doc. A/CONF.62/SR.136, A/CONF.62/SR.138, Manner, note 12 at 125–126.

confirmed the Declarations upon ratification.[38] Denmark made a corresponding declaration upon ratification in 2004.[39] Unlike Denmark, Finland and Sweden did not specify the conventions in question. The "long-standing convention" relevant to Finland and Sweden was the 1921 Geneva Convention relating to the Non-fortification and Neutralisation of the Åland Islands. Sweden's Declaration also referred to the 1857 Treaty. The Åland Strait and the Danish Straits could have been classified as straits where the right of transit passage applies without the Declarations of Finland and Sweden when they signed and ratified the 1982 UN Convention on the Law of the Sea and Denmark upon ratification in 2004.

Treaties covering the Turkish Strait and the Strait of Magellan clearly regulate navigation through the Straits. However, with respect to the Åland Strait and the Danish Straits some doubts remain, which have been expressed with regard to the applicability of the treaties according to the criteria of art. 35 (c). Some issues concerning the legal basis of 1857 Treaty as a convention were particularly reflected in art. 35 (c). The 1857 Danish Straits Treaty only referred to merchant vessels during times of peace but did not mention naval vessels because they did not pay Sound Dues. Some legal experts felt that no special regime for merchant vessels passing through the Danish Straits existed during peacetime because such vessels were governed by the right of passage, which was the general rule of international law at the time.[40] Sweden and Denmark firmly opposed this kind of interpretation. Sweden voiced its opinion in two Declarations made on the signature and ratification of the Convention. With regard to the Åland Strait the main question has typically related to geography because the Åland Convention only partly covers the Strait and it established a three nautical mile territorial sea around the Åland Islands.[41] The treaty arrangements of the Åland Islands have internationalized and neutralised the eastern

---

[38] "It is the understanding of the Government of Sweden that the exception from the transit passage régime in straits, provided for in Article 35 (c) of the Convention is applicable to the strait between Sweden and Denmark (Oresund) as well as to the strait between Sweden and Finland (the Aland islands). Since in both those straits the passage is regulated in whole or in part by long-standing international conventions in force, the present legal régime in the two straits will remain unchanged.", available at <http://treaties.un.org/pages/ViewDetailsIII. aspx?&src=UNTSONLINE&mtdsg_no=XXI~6&chapter=21&Temp=mtdsg3&lang=en#End Dec>

[39] "The Kingdom of Denmark makes the following declaration: It is the position of the Government of the Kingdom of Denmark that the exception from the transit passage regime provided for in article 35 (c) of the Convention applies to the specific regime in the Danish straits (the Great Belt, the Little Belt and the Danish part of the Sound), which has developed on the basis of the Copenhagen Treaty of 1857. The present legal regime of the Danish straits will therefore remain unchanged."

[40] E. Brüel *International Straits, A Treatise on International Law, Vol. II., Straits Comprised by positive Regulations* (Sweet and Maxwell, London, 1947) 45; C.J. Colombos *The International Law of the Sea 6th* (Longmans, 1967) 199; Manner, note 12 at 127, see also Alexandersson, note 2 at 73.

[41] J.A. Roach and R.W. Smith *Excessive Maritime Claims* 3rd (Martinus Nijhoff Publishers, 2012) 284.

part of the Åland Strait. Due to the Strait's international status and the provisions of the 1921 Åland Convention, certain limitations on the entry of warships into the zone have been set and the overflight of foreign military aircraft has been prohibited.[42] Furthermore, when taking into consideration the fact that these limitations focus partly on Finland as a coastal state, there seems to be no doubt that treaty arrangements of the Åland Islands regulate passage through the Åland Strait according to art. 35 (c).[43] By restricting Finnish naval vessels' access to sea areas surrounding the islands that are covered by the strait area, the 1921 Åland Convention regulates passage in part of the Åland Strait.

As a result of the application of the new concept of transit passage, however, the legal regime of straits would have changed. Had this happened, the regime of transit passage would have also had an influence on the political and military status quo within the Baltic region. The concept of transit passage would enable foreign submarines to navigate submerged through the Strait, for example, as well as the freedom of overflight for foreign aircraft. From the middle of the 1950s up to the beginning of the 1990s there were two military blocs, NATO and the Warsaw Pact Organization, which were active in the Baltic region. Both blocs had a great number of naval vessels, which were engaged in a number of intensive naval activities within the Baltic Sea.[44] The number of warships and their activities illustrate the extent of the importance that the region played in the strategies of both blocs. It is obvious that if the concept of transit passage had been applied to passage through the Åland Strait and the Danish Straits then states bordering the straits would be restricted in their efforts to follow and control foreign military activities.[45] Hence, Denmark, Finland and Sweden had valid grounds during the Third Law of the Sea Conference to state that it was essential to their national security to exclude the application of transit passage through those straits.[46] In addition to warships, the legal regime of the Baltic Sea straits also affects the ability of states bordering the straits to act against ship-based security threats, such as terrorism, drugs, human trafficking and international organized crime.[47] By applying the right of innocent passage in the Baltic Sea straits Denmark, Finland and Sweden as a part of international community can better meet the requirements for maritime security.

---

[42] Although the right of innocent passage of warships is allowed.

[43] Manner, note 12 at 128–129.

[44] B. Johnson Theutenberg *Folkrätt och säkerhetspolitik* (Nordstedts, Uppsala, 1986) 198. For more on the numerous unlawful intrusions and so called "submarine incidents" see Johnson Theutenberg at 381–384. At the time both Finland and Sweden required prior notification from foreign warships.

[45] Manner, note 12 at 124–125.

[46] Johnson Theutenberg, note 44 at 198.

[47] K. Bangert "Belts and Sund" *Max Planck Encyclopedia of Public International Law* available at MPEPIL online/Belts and Sund <http://www.mpepil.com> para. 16.

3.4   The Major Baltic Sea Straits

The Danish Straits have had commercial, political and military relevance in the region over the course of the centuries. Despite a significant change in the means of transport, the Danish Straits are still the main passages between the Atlantic Ocean and the Baltic Sea. The Sound is still the shortest and busiest passage between the Baltic Sea and the Kattegat but due to the increasing size of vessels the Great Belt has become the most frequently used route for the passage of large vessels because it is the only deep-water strait among the Danish Straits.[48]

The Danish project to build a bridge across the Great Belt caused a dispute between Denmark and Finland early in the 1990s. In May 1991 Finland filed an application with the International Court of Justice against Denmark concerning the right of passage of oil-rigs through the Great Belt. Finland claimed that the Danish project to build a bridge over the Great Belt, although connecting two parts of Denmark's territory, violated the right of free passage between the Baltic Sea and the North Sea established in the relevant Conventions and customary international law.[49] Finland stated that the bridge across the Great Belt would prevent the passage of Finnish drilling rigs through the strait and that it would be contrary to international law. The dispute was settled and Finland agreed to withdraw its application in return for the Danes paying a sum of 90 million Danish Crowns. The sum was calculated on the basis of the foreseeable damage the bridge could cause Finnish interests.[50]

The Åland Strait has also been an important area within international navigation for centuries; the Convention relating to the Non-fortification of and the Neutralisation of the Åland Islands in 1921 created a special legal status for the strait by demilitarizing and neutralizing the eastern side of the Märket Reef. Further the 1921 Åland Convention mentions the right of passage of foreign warships. Art. 5 of the 1921 Åland Convention says that warships enjoy the right of innocent passage through the zone and, at the time the Convention was concluded, this regulation also covered the eastern side of the Strait. By mentioning warships the Convention reserves the right of innocent passage to warships and also at the same time restricts the legal regime of the right to innocent passage.

The Åland Strait and the Danish Straits are narrow and the waters are shallow and heavy traffic through the straits creates a risk of stranding, collision and pollution. All vessels passing through the Åland Strait and Danish Straits have to pass through the territorial waters of the coastal states. Consequently, foreign vessels are subject to either the rules of non-suspendable innocent passage or transit passage, depending on the applicable legal regime.

---

[48] Manner, note 12 at 122; B. A. Boczek *International law: a dictionary* (Scarecrow Press, Inc., 2005) 313.

[49] M. Koskenniemi and M. Lehto "Finland and the Law of the Sea" in T. Treves (ed) *The Law of the Sea. The European Union and Its Member States* (Martinus Nijhoff Publishers, 1997) 127–150 at 149.

[50] Ibid.

3.4.1 The Åland Strait

The legal status of the Åland Islands is a special case in international law. Åland is an area which is both demilitarised, neutralised, and enjoys wide autonomy under Finnish sovereign rule. The corridor between the Baltic Sea and the Gulf of Bothnia is an international strait referred to as the "Åland Strait" (in Finnish: *Ahvenanrauma*), located within the Åland Sea.[51] As already mentioned above, the geography is a focal question in the Åland Strait because the treaty arrangements only partly cover the Strait..[52] The borderline between Finland and Sweden is found within the strait on Märket Reef (MR). It is also important to mention that Märket is only partly demilitarised and neutral due to the fact that the Swedish half is outside of the Conventions. However, the demilitarised and neutralised Finnish part of the Strait covered by the conventions is adequate to the application of art. 35 (c) that refers to a strait "regulated in whole or in part".[53] The route through the strait traverses Swedish territorial sea and is located between the Understen islet on the Swedish side and the Märket Reef.[54] The breadth of the Åland Strait measured at its narrowest point is no more than 6 nautical miles. However, this is not relevant when the scope of art. 35 (c) is considered.

As the right of innocent passage applies to navigation through the Åland Strait, to some extent Finland and Sweden both have an opportunity to control navigation through their respective EEZs within the Gulf of Bothnia because coastal states have a right to prohibit and suspend passage on the grounds that passage is not innocent. Furthermore, the passage of warships through coastal states' internal waters requires prior authorisation. The application of transit passage would have expanded the navigational rights of foreign ships in the Åland Strait. The rationale behind the concept of transit passage was to retain freedom of passage and overflight in straits where high sea routes would disappear because of the introduction of twelve-nautical-mile territorial seas. Nevertheless, it was accepted that this new regime would also cover straits where the right of innocent passage had previously applied. Hence, the original meaning of transit passage was expanded and the concept was intended to become the main rule as well as soon becoming a rule of customary international law. During the Third Law of the Sea Conference Denmark, Finland and Sweden agreed that it was necessary to exclude straits like the Danish Straits

---

[51] The Åland Strait is a significant sea route. According to HELCOM publication Shipping Accidents in the Baltic Sea, in 2013 14,433 ships on the Åland West route and 1397 ships on the Åland East route crossed AIS fixed lines through the Åland Strait, HELCOM *Annual report on shipping accidents in the Baltic Sea in 2013* (2014), 3–7.

[52] Roach and Smith, note 41 at 284.

[53] H. Rotkirch "The Demilitarization and Neutralization of the Åland Islands: A Regime "in European Interests" Withstanding Changing Circumstances" (1986) 23 *Journal of Peace Research* 357–376 at 372; Manner, note 12 at 126–127.

[54] C. N. Gregory "The Neutralization of the Aaland Islands" (1923) 17 *American Journal of International Law* 65–76 at 65; Rotkirch, note 53 at 359; for more information about the Åland Islands see the official homepage http://www.aland.ax

and the Åland Strait from the regime of transit passage. It was clear to Finland from the outset that negotiations would touch on matters related to the Åland Strait, particularly concerning the likely effects that treaty arrangements may have on the legal regime of the Åland Strait.[55] Hence, the treaty arrangements that partly covered the Åland Strait fulfil the requirements of art. 35 (c) of the 1982 Law of the Sea Convention. In addition, passage through the Åland Strait falls within the scope of innocent passage, which has been verified by Finland and Sweden in their Declarations when signing and ratifying the Convention.

The straits that are governed by art. 35 (c) are not listed in the article exclusively, which is why different interpretations have not been avoidable. Although Finland and Sweden mentioned that the Åland Strait also belongs to the category of straits "in which passage is regulated in whole or in part by long-standing international conventions in force", in some publications specifically relating to such straits it has been argued that the right of transit passage does not apply to the Åland Strait.[56] Roach and Smith believe that the territorial sea of the Åland Islands extends only by three nautical miles from the low-water mark and therefore does not extend beyond the outer limits of the straight line segments established by the 1921 Åland Convention. For this reason, the Convention is not applicable to the remaining waters that form the international strait.

Conventions concerning the legal status of the Åland Islands signify a long-standing legal tradition.[57] As the Commission of Jurists in 1920 had already mentioned in its statement,[58] the legal status of the Åland Islands was created with the best interests of European states in mind and therefore the provisions of the Convention were part of "European law". The 1856 Convention was complemented by the 1921 Åland Convention, which was supplemented by the 1940 Treaty.[59] In 1921 the Åland Convention concluded that it is complementary to the 1856 Convention and, moreover, it was stated that the 1921 Åland Convention is supplemented by the 1940 Treaty. The Second World War also influenced the legal status of the Åland Islands. Furthermore, the 1947 Peace Treaty confirmed the status quo

---

[55] UNCLOS III, Official Records, Volume XIV, 21, 35; Hakapää, note 18 at 202.

[56] W.L. Schachte Jr. and J.P.A. Bernhardt "International Straits and Navigational Freedoms" (1992–1993) 33 *Virginia Journal of International Law* 527–556 at 547–548; *Limits in the Seas No. 112*, 67; Roach and Smith, note 41 at 284.

[57] France, Great Britain and Russia concluded the Convention on the demilitarisation of the Åland Islands in 1856 after the Crimean War; the 1921 Convention on the Demilitarisation and Neutralisation of the Åland Islands; the demilitarisation of the Åland Islands is also mentioned in the 1947 Paris Peace Treaty, whereby Finland and the Soviet Union concluded the bilateral treaty concerning the demilitarisation of the Åland Islands in 1940.

[58] Report of the International Committee of Jurists 1920, 17–19.

[59] Manner, note 12 at 127–128; L. Hannikainen "The Continued Validity of the Demilitarised and Neutralised Status of the Åland Islands" (1994) 54 *Zeitschrift für ausländisches öffentliches Recht und Völkerrecht* 614–651 at 619–620.

by declaring that "the Åland Islands will remain demilitarised according to the present situation". Hence, the political climate that existed before the war was retained.[60]

The convention that had endorsed the application of art. 35 (c) was the 1921 Convention. The provisions of the 1921 Åland Convention set obligations for Finland to limit the entrance or stay of foreign warships within the zone defined by the Convention. However, the right of innocent passage through Åland's territorial waters is not restricted. Nevertheless, the sea route through the strait is located within the territorial waters of Sweden. Therefore, passage is regulated by Swedish legislation and not the 1921 Åland Convention.[61] However, east of Märket Reef the Strait belongs to the demilitarised and neutralised area and also belongs to Finland's internal waters. Furthermore, Finland's territorial sea and internal waters cover the sea area north and south of the Märket Reef and part of these areas also belong to the demilitarised and neutralised zone where the right of innocent passage applies. There is no doubt about the application of the 1921 Åland Convention in the strait area belonging to Finland and therefore the provisions of the 1921 Convention regulating the entrance of warships to the archipelago are also applied in the Finnish part of the strait.

Art. 35 (c) states that the Strait "is regulated in whole or in part by long-Standing international conventions". The article does not set any requirements for a part of a Strait that is covered by long-standing conventions. Finland and Sweden have both interpreted the Åland Conventions to be a long-standing convention, which means that the Finnish part of the strait is governed by art. 35 (c). Other states did not question this position during the Conference; not even the United States.

This exception to the rules of passage through straits regulated by long-standing international conventions was very important to Finland and Sweden because without it the *status quo* in that part of the Baltic Sea would have been altered.[62] Any application of the concept of transit passage to passage through the Åland Strait would have nullified limitations related to the passage of submarines and the overflight of aircraft. Submarines would then be allowed to navigate submerged through the Strait if the right of transit passage would have been applied instead of the right of innocent passage. Transit passage would have also contributed to the overflight of military aircraft because the right of innocent passage does not include the right of overflight, which the right of transit passage does.[63]

---

[60] B. Broms *Kansainvälinen oikeus* (Suomalainen lakimiesyhdistys, Vammala, 1978) 534; Johnson Theutenberg, note 44 at 203; Rotkirch, note 53 at 372.

[61] Foreign state ships have a right of passage through the Swedish territorial sea. Passage may not be prejudicial to the peace, good order or security of Sweden and must be continuous and expeditious. Ordinance 1992:118 (Admission Ordinance) Permission to enter Swedish airspace is required for a foreign state aircraft, including military helicopters.

[62] Rotkirch, note 53 at 372. At the time of the Conference, Finland and Sweden required prior notification for the passage of foreign warships through their respective territorial seas, whereas this kind of requirement was not possible when exercising transit passage.

[63] M. Lehto *Itämeren turvallisuusjärjestelmä erityisesti oikeudellisen säännöstön kehityksen kannalta* (ARNEK, 1986) 56.

However, the 1921 Åland Convention only covers parts of the Strait belonging to Finland. It is noteworthy that the sea route passes through the Swedish side of the strait as well. However, when the legal status of the Åland islands is examined one must bear in mind that art. 35 (a) concerning the internal waters is important because the eastern side of the Märket Reef belongs to the internal waters of Finland. As a general rule, those areas of straits used for international navigation, which are also a part of internal waters, preserve their status and the 1982 Convention provisions on the special regime of straits are not applied. Nevertheless, in circumstances where parts of the seas form a part of a coastal state's territorial sea area, EEZ, or high seas and because of the use of straight baselines, the sea areas become coastal state's internal waters and the right of innocent passage or transit passage is retained in accordance with the provisions of the 1982 Convention. However, Finland established its straight baselines before the law of the sea conventions came into force, which were based on customary international law and did not provide that passage rights were retained. Consequently, it seems that any exception regarding the right of innocent passage or transit passage through Finland's internal waters is not applicable to the eastern side of the Märket Reef. Thus, passage on the eastern side of Märket Reef cannot take place without the permission of Finnish authorities. On the other hand, it seems that in light of the 1921 Åland Convention and as a consequence of the establishment of straight baselines, the sea area where stricter restrictions are applied to the navigation of Finnish and foreign warships is expanded.

3.4.2   The Danish Straits

As early as 1429 the Danish government began collecting a transit duty on vessels passing through the Sound. In response to various protests made by those who questioned the legality of the impost, King Erik of Pomerania claimed that the straits were part of Danish territory and that the Baltic Sea was a private sea area that fell under Danish sovereignty. Transit through the Belt was prohibited until the end of the fifteenth century when a similar duty was imposed on vessels permitted to sail through the Great Belt.[64]

There was much negative criticism of the Sound dues in the early 1820s as mainly foreign governments but also merchants from Copenhagen expressed their sentiment on the issue. The Danish government was naturally unwilling to abolish the dues because of the implications for the state budget. Finally, an announcement was made in 1845 when the United States refused to pay the Sound dues. The United States questioned Denmark's right to collect the dues "under the public law of nations". However, a rapid growth in commerce and shipping in the United States

---

[64] Brüel, note 40 at 33; Alexandersson, note 2 at 70; S.P. Menefee "The Sound Dues and Access to the Baltic Sea" in R. Platzöder and P. Verlaan (eds) *The Baltic Sea: New Developments in National Policies and International Cooperation* (Martinus Nijhoff Publishers, 1996) 101–132 at 102–103.

hastened endeavors to remove all trade barriers.[65] The United States' opinion encouraged a general revision of the trade barrier issue in 1857, which led to the signing of two Conventions on the Sound and the Belts.[66] The Copenhagen Convention on the Sound and the Belts was signed by European naval states and the Washington Convention was signed by Denmark and the United States. According to both Conventions, Denmark would receive compensation if it allowed foreign ships "free passage". The dues were finally discontinued after 1857 as a result of the Treaties. Since the 1857 Treaties no other multilateral treaties or conventions have been concluded with regard to the Danish Straits. However, art. 195 of the Treaty of Versailles mentions the right of "free passage into the Baltic to all nations". The Sound Dues Conventions only applied to merchant vessels in the Danish Straits as naval vessels were not covered by the 1857 Treaty because they did not pay Sound Dues.[67]

Other questions concerning the Danish Straits stemmed from the use of the term "free passage", which was mentioned in the 1857 Treaties. In the Conventions this term meant freedom from the Sound Dues. The Sound and the Great Belt are parts of the territorial seas belonging to states bordering the straits, which means that navigation through these waters must comply with the rules of innocent passage. In the case regarding passage through the Great Belt, Denmark stated that the rules of international law applicable to the right of innocent passage through the Danish Straits are the Treaty of 1857 for the Redemption of the Sound Dues and the 1958 Convention on the Territorial Sea and Contiguous Zone.[68] The 1857 Treaty of Copenhagen says that "[n]o vessel shall henceforth, under any pretext whatsoever, be subjected, in its passage of the Sound or Belts, to any detention or hindrance".[69] However, Denmark and Sweden regulate the right of passage through the Danish Straits through their domestic legislation.[70] The Danish legislation, Ordinance Governing the Admission of Foreign Warships and Military Aircraft to Danish Territory in Time of Peace, regulates the approach to the Baltic Sea. Simultaneous passage through the Great Belt, Samsoe Belt or the Sound of more than three warships of the same state is subject to prior notification. It has been stated by some writers that coastal state domestic legislation is not in contradiction with provisions

---

[65] Brüel, note 40 at 36–37.

[66] See Treaty between Great Britain, Austria, Belgium, France, Hanover, Mecklenburg-Schwerin, the Netherlands, Oldenburg, Prussia, Russia and Sweden, and Norway and the Hansa Towns on the one hand, and with Denmark on the other, for the Redemption of the Sound Dues, signed at Copenhagen and Convention between United States of America and Denmark for the Discontinuance of the Sound Dues 116 PCTS 357.

[67] Alexandersson, note 2 at 72–73; Manner, note 12 at 12.

[68] Passage through the Great Belt (Finland v. Denmark), Counter-Memorial Submitted by the Government of the Kingdom of Denmark, 219; Manner, note 12 at 123–124.

[69] Churchill and Lowe, note 28 at 114; S. Mahmoudi "The Baltic Straits" in D. D. Caron and N. Oral (eds) Navigating Straits. Challenges for International Law (Brill/Nijhoff, 2014) 125–131 at 130.

[70] Note 61. Ordinance Governing the Admission of Foreign Warships and Military Aircraft to Danish Territory in Time of Peace 224, 16 April 1999.

of the 1857 Treaty nor arts. 21–23 of UNCLOS, but with respect to enforcement jurisdiction it seems that a coastal state's legislative competence is not applicable.[71] Thereby without the exception provided for in art. 35(c) the legal regime of the Danish Straits would have changed. This would have meant that the new rules would have invalidated by the old treaties establishing particular regimes for Danish Straits. In these cases, the Parties to the Conventions would have been obliged to approve and comply with the provisions of UNCLOS.

According to the US Department of State publication, Limits in the Seas No. 112, the United States position is that passage of warships and state aircraft through the Danish Straits is based either upon the customary right of transit passage or upon the conventional right of "free and unencumbered navigation".[72]

3.5 Interpretive Positions Regarding Art. 35 (c)

The concept of transit passage was not created in order to merely expand the right of transit passage through straits used for international navigation but was instead intended to maintain freedom of navigation in spite of the widening of territorial seas, as this tended to lead to a disappearance of high sea corridors where warships and military aircraft had previously enjoyed freedom of navigation and overflight. However, although the initial purpose of the regime of transit passage was to retain freedom of navigation through straits where high sea routes disappear as territorial seas expand, during the negotiation process of the Third Law of the Sea Conference it also emerged that the right of transit passage would cover straits where previous high sea routes had not existed. Therefore, the application of the concept of transit passage would have expanded the regime of transit passage through the Åland Strait and the Danish Straits, i.e. the provisions on transit passage would have affected the legal regime of straits regulated by the old conventions. The application of art. 35 (c) is ambiguous in the sense that it does not include a list of straits that it covers, which therefore leaves scope for various interpretations.

Declarations issued by Finland and Sweden as well as Denmark state that it is sufficient that part of the strait is governed by a long-standing treaty already in force. The wording of art. 35 (c) is ambiguous in the sense that it does not clearly define the precise area where this kind of treaty is applied. Interpretations by Finland, Sweden and Denmark seem to suggest an extension of the legal influence of such a treaty to vessels and to parts of a Strait that are beyond the scope of the treaty. Thus, passage through the Åland Strait and the Danish Straits is governed by the rules relating to innocent passage, which is not allowed to be suspended by states bordering the strait.

The term used in the text, "in whole or in part" can be interpreted in several ways. Schachte and Bernhardt present some of the possible interpretations of the term.

---

[71] Churchill and Lowe, note 28 at 114; Mahmoudi, note 69 at 131.

[72] See *Limits in the Seas No. 112*, 75.

Firstly if a convention governs a given strait but only according to certain circumstances, for instance with regard to merchant vessels but not aeroplanes, because art. 35 states from the beginning that "nothing in this part affects: the legal regime in straits", then straits must be regulated by other rules of customary international law.[73] However, if a convention were to regulate a strait partly, in the sense that one part fell under provisions of art. 35 (c) and the other part remained unregulated, then the right of transit passage would apply. Schachte and Bernhardt believe that, due to the fact that transit passage is the norm, but includes an exception as well as circumstances where the exceptional regime does not cover every aspect, the normal regime should be used to fill in the gaps.[74]

As noted by Schachte and Bernhard a strait may have a different legal status depending on a vessel's character. Therefore, when passage of naval vessels takes place within certain straits then the right of transit passage must be applied, whereas the passage of merchant vessels is regulated by long-standing Conventions. It is highly probable that some form of confusion regarding the correct Convention to use could have occurred when states agreed upon the exceptions to the legal regime. It is even more likely that states had intended for long-standing Conventions to be sufficient enough when seeking to regulate different straits because the Conventions alluded to a number of alternative solutions proposed during negotiations. The legal statuses of straits are permanent and do not vary according to the types of vessels or other factors like geography.

The US Department of State has expressed in the publication Limits in the Seas that art. 35 (c) is not applicable to the Danish Straits. The 1857 Danish Straits Treaty only referred to merchant vessels during times of peace but did not mention naval vessels because they did not pay Sound Dues. The United States position is that passage of warships and state aircraft through the Danish Straits is based on either the customary right of transit passage or the conventional right of "free and unencumbered navigation".[75]

Although Finland and Sweden mentioned during negotiations that the Åland Strait also belonged to the category of straits "in which passage is regulated in whole or in part by long-standing international conventions in force", in some publications it has been argued that the right of transit passage does not apply in the Åland Strait.

Rauch states that there is no specific regulation of passage through the Åland Strait because the 1921 Åland Convention does not restrict the right of innocent passage of warships through the demilitarised and neutralised zone and as result, such passage remains subject to existing international rules and usages.[76] Art. 5 of

---

[73] See art. 35 (c); Schachte and Bernhardt, note 56 at 544.

[74] Schachte and Bernhardt, note 56 at 544–545.

[75] *Limits in the Seas No. 112*, 75.

[76] E. Rauch *The Protocol Additional to the Geneva Conventions for the Protection of victims of International Armed Conflicts and the United Nations Convention on the Law of the Sea: Repercussions of Naval Warfare. Report to the Committee for the Protection of Human Life in Armed Conflict of the International Society for Military Law and Law of War* (1984) 52.

the 1921 Åland Convention specifically says that: "The prohibition to warships of entering and remaining in the zone described in art. 2 does not restrict the freedom of innocent passage through the territorial waters. Such passage shall remain subject to existing international rules and usages." Furthermore the Convention prohibits the overflight of foreign military aircraft. The phrasing of art. 5 reflects the possibility that if rules regarding warships' right of innocent passage were to change then it would essentially affect the application of this article. The right of innocent passage is the only applicable legal regime of passage for foreign warships through the demilitarised and neutralised sea area around the Åland archipelago.

Roach and Smith as well as Heintschel von Heinegg believe that the territorial sea of the Åland Islands extends only by three nautical miles from the low-water mark and therefore does not extend beyond the outer limits of the straight line segments established by the 1921 Åland Convention. For this reason, the Convention is not applicable to the remaining waters that form the international strait.[77]

As regards the Åland Strait, the US Department of State has expressed that art. 35 (c) is not applicable to the Åland Strait because the territorial sea of the Åland Islands only extends by 3 nautical miles from the low-water line..[78]

However, the official documents of the conference do not support the above-mentioned interpretations relating to the Åland Strait and Danish Straits. Negotiations held between the delegations of the naval powers (the United States, the Soviet Union and Great Britain) on the one hand and the delegations of Denmark, Finland and Sweden on the other, were difficult but a consensus was reached. Furthermore, no state has officially protested against the Finnish and Swedish interpretation stated in their declarations upon the signing and ratification of the Convention.[79]

According to documents prepared by members of the Finnish delegation to the Conference relating to the Åland Strait, Finland held negotiations with the Soviet Union after the Caracas session and with the United States in Geneva. Both the United States and the Soviet Union separately informed the Finnish delegation that they interpreted the draft article in a way that the nature of passage in the Åland Strait was not going to change and the right of innocent passage would continue to be applied to passage through the Åland Strait.[80]

---

[77] W. Heintschel von Heinegg "The Law of Naval Warfare and International Straits" in M.N. Schmitt and L.C. Green (eds) *The Law of Armed Conflict: Into Next Millennium* (71 International Law Studies, 1998) 263–292 at 279; Roach and Smith, note 41 at 284.

[78] *Limits in the Seas No. 112*, 66–67: "The United States, which is not a party to this Convention, has never recognized this international strait as falling within the Article 35(c) exception."

[79] K. Hakapää, J. Perttunen and H. Rotkirch *YK:n III merioikeuskonferenssi. Selostus 2. istuntokaudesta Garacasissa 20.6–29.8.1974* (1975) 112; K. Hakapää, T. Hämäläinen, E. Kivimäki and T. Lahelma, H. Rotkirch *YK:n III merioikeuskonferenssi. Selostus 3. istuntokaudesta Genevessä 17.3–10.5.1975* (1975) 86; K. Hakapää "National Interests and Policies of Finland in the Baltic Sea: A Law of the Sea Perspective" in R. Platzöder and P. Verlaan (eds) *The Baltic Sea: New Developments in National Policies and International Cooperation.* (Martinus Nijhoff Publishers, 1996) 387–398 at 397; Koskenniemi and Lehto, note 49 at 133.

[80] Hakapää, Hämäläinen, Kivimäki, Lahelma and Rotkirch, note 79 at 86–88.

With regard to passage through this zone, Finnish warships enjoy same rights as foreign states' warships which exercise their right of innocent passage. These restrictions should be seen as exceptions to the doctrine of sovereignty, which a coastal state would normally exercise in its territorial sea. Finland's sovereign rights as a coastal state are significantly restricted by the 1921 Åland Convention.

The definition of a strait does not limit either the length or the breadth of any strait and it can be said that only the narrowest point of the Åland Strait belongs to Finland's internal waters. However, the right of innocent passage exists in those parts of the strait that became internal waters after the establishment of the straight baselines in 1995 because of art. 8(2) of UNCLOS and it also applies to passage through the strait in areas that belong to the demilitarised and neutralised zone. Thus as art. 5 of the 1921 Convention only speaks about the right of innocent passage of warships it can be said that the passage through the Åland Strait is regulated, in part, by the long-standing Convention in force.

It should be borne in mind that the wording of art. 35(c) is a compromise and that art. 35(c) is part of the "package deal". It was the intention of the negotiations at the Conference to bring the Åland Strait and the Danish Straits under this provision.[81]

3.6 Conclusions

This paper has aimed to demonstrate that because the straits that are governed by art. 35 (c) are not listed in the article individually, it allows for an interpretation that was not expressed during the Third Law of the Sea Conference. Art. 35 (c) has not resolved the problems that emerged in the connection with "transit passage", a new legal regime for straits used for international navigation. The declaration given by Finland, Denmark and Sweden implies that the Åland Strait and the Danish Straits have a special regime based on the long-standing international conventions. Therefore, any changes in the rules of international law based on treaties or customary international law applicable to straits in general are not applicable to passage through the Åland Strait or Danish Straits. There are a number of different interpretations of the legal regime of the Åland Strait and the Danish Strait, which implies that the exact content of the legal regime of these straits is not clear. The straits that are governed by art. 35 (c) are not listed in the article, however, on the base of discussions in several meetings held at the Third Law of the Sea Conference, it seems that art. 35 (c) applies not just to the Turkish Straits and the Strait of Magellan, but also to the Åland Strait and the Danish Straits. It is also noteworthy that other states did not question the position Finland, Sweden and Denmark during the Conference.

Passage through the Åland Strait and the Danish Straits is subject to the regime of innocent passage which means that passage is non-suspendable, although there exists more restrictions for ships in the passage than under a regime of transit passage. If the regime of transit passage were to have been applied to the Åland Strait

---

[81] Menefee, note 64 at 129.

and the Danish Straits it would have revolutionized navigation in the Baltic Sea because submarines would have been able to navigate, submerged, to the Baltic Sea through the Danish Straits and through the Åland Strait towards the Gulf of Bothnia, as well as through the northern strait called Kvarken (in Finnish: *Merenkurkku*). Therefore, foreign submarines could navigate continuously through the Baltic Sea until reaching the Gulf of Bothnia while being submerged. If this were to have happened it would have certainly upset the military balance in the Baltic Region because when the Convention was signed, the region was greatly influenced by two military blocs. If transit passage is applied in the Åland Strait it might be necessary to reconsider the maritime zones in the northern strait. The effects of the application of the transit passage regime might be in opposition to its ultimate purpose in the Baltic Sea region not to change navigational freedoms through the straits because that could lead to instability and tension. The key issue is balancing the interests of the Baltic Sea coastal states and the third states and one way to do this is not to change passage rights through the Åland Straits and the Danish Straits and thus maintain the right of innocent passage in these straits. The right of innocent passage grants, for foreign ships, non-suspendable access to and from the Baltic Sea and Gulf of Bothnia and balances the interests of those states that border the straits with the requirements for international maritime communication.

Chapter 4
Regulating Eutrophication – Flexible Legal Approaches and Environmental Governance in the Baltic Sea Area

Brita Bohman

Abstract The chapter discusses the design of regulatory instruments and structures adopted to deal with eutrophication in the Baltic Sea area. New forms of regulations have been introduced in the past decades and what may appear as regulatory gaps or uncertain obligations is often closely connected to the developing idea of an ecosystem approach to regulation, in international as well as EU law. The chapter assesses what this concept actually entails in legal terms and whether the demand for regulation and enforcement of measures to reduce the eutrophication is sufficiently and effectively covered in the existing regulatory framework. It also reviews benefits and obstacles linked to ecosystem-based regulation and assesses whether the concept provides is effective in abating environmental problems.

Keywords Eutrophication · Baltic Sea · Ecosystem approach · Good environmental status · Helsinki convention · Baltic Sea action plan · Marine strategy framework directive · Baltic Sea

4.1 Introduction

The topic in this chapter is legal structures and approaches to regulating eutrophication in the Baltic Sea area. The problem of eutrophication is one of the major environmental problems in the Baltic Sea. It shares similar features with other complex environmental problems. Finding effective regulatory approaches might thus also serve a purpose in other marine areas. Eutrophication is the result of an overload of nutrient pollution (phosphorus and nitrogen) from a variety of sources in the catchment area such as industries, water management plants, animal farms, agriculture,

B. Bohman (✉)
School of Business, Economics and Law, Department of Law,
University of Gothenburg, Sweden
e-mail: brita.bohman@law.gu.se

© Springer International Publishing AG, part of Springer Nature 2018
H. Ringbom (ed.), *Regulatory Gaps in Baltic Sea Governance*, MARE
Publication Series 18, https://doi.org/10.1007/978-3-319-75070-5_4

45

and forestry.[1] The problem is complicated to abate due to factors like the sensitive nature of the marine ecosystem, delayed environmental response to pollution reductions, as well as the diversity and nature of the pollution sources. The sources of nutrients are to large extent diffuse and, compared to point sources, hard to trace, monitor, or reduce directly.

As eutrophication is a central environmental concern in the Baltic Sea area, it is addressed by a number of legal instruments. Since the 1970s it has primarily been regulated through the Helsinki Convention[2] and the Baltic Marine Environment Protection Commission – the Helsinki Commission (HELCOM). In recent years, other significant regulatory instruments have also been adopted. An instrument directly related to the Helsinki Convention is the HELCOM Baltic Sea Action Plan (BSAP) from 2007. The BSAP has uncertain legal status in comparison to other instruments applicable, but it still has an important impact on the regulatory situation. A number of legal instruments adopted by the European Union (EU) are also applicable, primarily the EU Marine Strategy Framework Directive (MSFD)[3] from 2008. The MSFD is applicable to marine areas but should be assessed in combination with the EU Water Framework Directive (WFD),[4] which is applicable to inland waters and as part of that regulate activities with indirect effect on marine areas. A range of area specific directives are furthermore applicable. Most of them are included within the frame of the WFD and they are thereby part of the operationalization of both the WFD and the MSFD.[5]

The MSFD creates a direct connection between the EU regulations and the international regulation through an obligation for the member states to use: "…[E]xisting regional institutional cooperation structures, including those under Regional Sea Conventions, covering that marine region or sub-region…", in order to achieve coordination in the implementation.[6] The BSAP was adopted with the intention to

---

[1] This paper focuses primarily on land-based waterborne sources, which are the most significant sources, although also air-borne pollution and pollution from *inter alia* maritime transport also contributes to the total pollution. On the causes of eutrophication and its main sources see *inter alia*; R. Elmgren "Understanding Human Impact on the Baltic Ecosystem: Changing Views in Recent Decades" (2001), Vol. 30, No 4/5, *Ambio*, 222–231; HELCOM (2015) Baltic Sea Environment Proceedings No. 145, "Updated Fifth Baltic Sea pollution load compilation (PLC-5.5)" at 25.

[2] The Convention on the Protection of the Marine Environment of the Baltic Sea (adopted 9 April 1992) 2099 UNTS 195. The Helsinki Convention was first adopted in 1974 but was revised in 1992. The current version of the Helsinki Convention is thus the 1992 Helsinki Convention and has slightly different provisions. For a review of the difference between the original and the revised version of the Helsinki Convention, see e.g. J. Ebbesson "A Critical Assessment of the 1992 Baltic Sea Convention" (2000) 43 *German Yearbook of International Law*, 38–64.

[3] Directive 2008/56/EC, OJ L 164, 25.6.2008, 19–40.

[4] Directive 2000/60/EC, OJ L 327, 22.12.2000, 1–73.

[5] Area specific directives with significance for eutrophication are *inter alia*: Urban Waste Water Treatment Directive (91/271/EEC), OJ L 135, 30.5.1991, 40–52; The Nitrates Directive (91/676/EEC), OJ L 375, 31.12.1991, 1–8, and; The Industrial Emissions Directive (2010/75/EU), OJ L 334, 17.12.2010, 17–119.

[6] The MSFD, art. 6(1).

create an instrument for regional implementation of the MSFD in the Baltic Sea marine region.[7] This regulatory construction creates a complex structure of over-lapping and intertwined regulations, which also affects interpretation, implementation and application of the legal instruments. It furthermore puts the BSAP with its uncertain legal status in a central position.[8] The extent of the consequences and results of this interaction is, however, unclear due to the fact that all instruments have an open and rather vague design.

The Helsinki Convention is a typical international agreement. It includes some rather flexible requirements. It is also vague mainly due to the compromises that are often the result of international negotiations.[9] The BSAP and the MSFD (as well as the WFD), as a contrast, are intentionally flexible and integrating a goal-oriented approach to regulation with a base in the concept of ecosystem approach.[10] One main factor in this approach is that the environmental status is in the center of the regulatory design chosen for these instruments. Beyond flexibility this also entails an adaptive approach, since circumstances will change or be changed with the fluctuations of the environmental factors that steers the system. The ecosystem approach is a complicated concept that generally proposes a closer integration of environmental governance theories into legal structures.[11] The environmental or ecological status is hence meant to steer the choice of measures and actions taken in the implementation. A wide leeway is provided for the states to decide. The MSFD and the BSAP has set up a structure for monitoring measures and related actions, which imply that the measures taken shall be evaluated in relation to environmental changes and ecological indicators defined for this.[12] This legal design entails that

---

[7] This is clear from the statements made in the Joint HELCOM-OSPAR Ministerial Meeting, held in Bremen 2003, see HELCOM (2003a), JMM 2003/3(final version)-E, agenda item 6. "Declaration of the Joint Ministerial Meeting (JMM) of the Helsinki and OSPAR Commissions". The aim to coordinate the work of the BSAP with the EU MSFD is also made clear in the preamble of the BSAP.

[8] See *inter alia*, J. Van Leeuwen, L. Van Hoof and J. Van Tatenhove "Institutional ambiguity in implementing the European Union Marine Strategy Framework Directive" (2012) 36 *Marine Policy* 636–643; R. Barnes and D. Metcalfe "Current Legal Developments – The European Union – The Marine Strategy Framework Directive" (2010) 25 *The International Journal of Marine and Coastal Law* 81–91, at 86–88.

[9] See e.g. T. Gehring "Treaty-Making and Treaty Evolution" in D. Bodansky, J. Brunneé and E. Hey (eds.) *The Oxford Handbook of International Environmental Law* (Oxford University Press, Oxford, 2007) 467–497, at 470.

[10] See further elaborations on the background of this concept in the section The Ecosystem approach in this chapter.

[11] See e.g. V. De Luca "Competing Narratives and Complex Genealogies: the Ecosystem Approach in International Environmental Law" (2015) 27 *Journal of Environmental Law* 91–117.

[12] The MSFD arts. 1 and 5. See also *inter alia* R. Barnes and D. Metcalfe "Current Legal Developments – The European Union – The Marine Strategy Framework Directive" (2010) 25 *The International Journal of Marine and Coastal Law* 81–91, at 82–85; À. Borja, M. Elliott, J. Carstensen, A-S. Heiskanen and W. Van de Bund "Marine Management – Towards and integrated implementation of the European Marine Strategy Framework and the Water Framework Directives" (2010) 60 *Marine Pollution Bulletin* 2175–2186 and; N. Zampoukas, H. Piha, E. Bigagli, N. Hoepffner, G. Hanke and A. C. Cardoso "Marine monitoring in the European Union:

while eutrophication in the Baltic Sea is a highly regulated issue area with overlapping regulatory instruments, there are still regulatory gaps and uncertainty remaining. This uncertainty is connected to the fact that much detail of indicators for environmental status, required measures, and obligations are left to be deemed and decided by the states in their individual assessments. In addition it is still rather unclear what an ecosystem approach implies in a regulatory setting. The combination of different regulatory regimes and their inter-connectedness also contributes to a blur of what each instrument actually requires and to what extent their obligations could be considered applicable.

4.2   Purpose, Aim and Outline

The purpose of this chapter is to discuss and review the design of the regulatory instruments and structures adopted to abate the eutrophication in the Baltic Sea area. What may appear as regulatory gaps or uncertain obligations within these regulatory instruments is to large extent connected to the ambition to integrate an ecosystem approach. The question is whether demand for regulation and legal enforcement of measures to reduce the eutrophication is sufficiently and effectively covered by these structures and approaches. It is also a question of general concern to investigate what the implementation of an ecosystem approach actually entails in legal context. The aim of the chapter is to investigate what these regulatory instruments and the integrated ecosystem approach entail in terms of requirements, control, enforcement and goal-achievement. The aim is furthermore to provide a review of benefits and potential obstacles connected with regulatory arrangements that are based on an ecosystem approach, and assess whether the concept of ecosystem approach provides a basis for further effectiveness in regards to abating environmental problems. In order to discuss these issues the following questions will be in focus:

- What is the concept of ecosystem approach and what does the concept add to the regulatory design of complex environmental problems? What does the perspective of environmental governance connected to the ecosystem approach imply in a regulatory setting?
- What does the flexible regulatory approach, which follows from integrating the ecosystem approach, entail in terms of control, enforcement, and goal-achievement? What does it entail more specifically for the regulation of eutrophication in the Baltic Sea area?
- Is the legal nature equipped to meet and somehow bridge the uncertainty that might arise in the combination of law and governance that are connected to the ecosystem approach?

How to fulfill the requirements for the marine strategy framework directive in an efficient and integrated way, Short Communication" (2013) 39 *Marine Policy* 349–351.

4.3 The Ecosystem Approach

As stated in the introduction, both the BSAP and the MSFD are explicitly designed to integrate an ecosystem approach in the regulatory structures of the Baltic Sea area. It is a principal point in assessing these regulatory structures in relation to abating eutrophication. It is therefore interesting to further elaborate on this concept and what it entails for regulation. The concept of ecosystem approach was formally accepted when adopted in the Convention of Biological Diversity[13] (CBD) in 1995.[14] The interpretation and application of the concept has since then been elaborated in guidelines and principles developed by the CBD.[15] These elaborations and interpretations are important for understanding the concept also in other contexts. The guidelines and principles developed as part of the work in the CBD also provide a basis for understanding of the concept as adopted by HELCOM. HELCOM has even referred directly to the CBD in the background documents to its own implementation and integration of the concept.[16]

In summary the management principles suggest that an ecosystem approach should be based on integrated and adaptive management, adjusted to deal with environmental complexity and ecological dynamics also in the absence of complete knowledge. Measures should be applied at a compatible level with the ecosystem and the concept holds *inter alia* stakeholder participation as valuable for such management approach.[17] More precisely what this entail in legal terms and measures is however nowhere defined or clarified. Management with lack of complete knowledge is especially relevant in relation to the eutrophication of the Baltic Sea since eutrophication to large extent is characterized by complexity in terms of non-linearity and time-lags. The issue of measures in absence of complete knowledge could also be connected to the precautionary principle. But in the context of ecosystem approach the precautionary principle might have a different interpretation or at least different prerequisites for application due to the adaptive and integrated

---

[13] The Convention on Biological Diversity (adopted 5 June 1992) 1760 UNTS 79.

[14] CBD (1995) CBD II/8 "The Second Meeting of the Conference of the Parties (COP) of the Convention on Biological Diversity. Decision 8." UNEP/CBD/COP/2/19, November 1995, at 12.

[15] Description and principles of the Ecosystem Approach were presented to the CBD (2000a) COP5 as SBSTTA 5 Recommendation V/10, January/February 2000.. available at <www.cbd.int/ recommendation/sbstta/default.shtml?id=7027>. Further elaborations has also been made, e.g. in 2004 at the CBD (2004) COP 7, in decision VII/11, UNEP/CBD/COP/DEC/VII/11, April 2004.

[16] HELCOM (2003b) "Statement on the Ecosystem Approach to the Management of Human Activities 'Towards and Ecosystem Approach to the Management of Human Activities'", Document adopted at the First Joint Ministerial Meeting of the Helsinki and OSPAR Commissions (JMM) Bremen, 25–26 June 2003, Agenda ítem 6, Annex 5, (Ref. §6.1).

[17] CBD (2000b) CBD V/6, section A, para. 4; According to the CBD, the ecosystem approach "… requires adaptive management to deal with the complex and dynamic nature of ecosystems and the absence of complete knowledge or understanding of their functioning. Ecosystem processes are often non-linear, and the outcome of such processes often shows time-lags."

approach entailed.[18] It is also connected to the fact that the ecosystem approach shall be based on relevant scientific methodologies with respect to ecosystem organization, process and function. Relevant management or regulatory measures taken should be based on science and have the ecosystem peculiarities in focus.

The CBD management principles are to large extent focused on capacity-building, local participatory involvement and continuous monitoring in multi-level institutional structures, while introducing also a number of basic points of departure for management.[19] In its origin this concept overlap to large extent with general strategies and theories of ecosystem management, and refers to a comprehensive science-based approach to the conservation and management of natural resources.[20] It has however successively also become a legal concept. The ecosystem approach is to be seen as a concept for law to integrate a specific focus on ecological prerequisites and environmental governance features, aiming to take into account the multi-leveled dynamics of regulatory and institutional pluralism typical of modern regulatory structures. This is an approach that to large extent draws on basic features of ecosystem adaptive management and governance ideas.[21] This is also characteristic of the difficulties found in applying the concept, since basic features of ecosystem management diverge significantly from basic features of legal systems.

The concept of ecosystem approach is an important step towards ecosystem-based legal measures, and holistic, dynamic perspectives in environmental regulation. It could add a tool for a more effective regulatory approach to environmental problems if its aim and foundation is properly transferred into law. However, it still remains to more concretely define how it should be expressed and transposed into legal rules and mechanisms.[22] The regulatory instruments of the Baltic Sea provide a view of how these management principles and approaches could be transposed into legal structures.

---

[18] See e.g.: A. Trouwborst "The Precautionary Principle and the Ecosystem Approach in International Law: Differences, Similarities and Linkages" (2009) 18(1) *Review of European Community and International Environmental Law* 26–37.

[19] See CBD (2004) CBD COP 7, decision VII/11, UNEP/CBD/COP/DEC/VII/11, April 2004.

[20] H. Wang "Ecosystem Management and Its Application to Large Marine Ecosystems: Science, Law, and Politics" (2004) 53(1) *Ocean Development and International Law* 41–74, at 43.

[21] See e.g. H. Österblom et al. "Making the ecosystem approach operational—Can regime shifts in ecological- and governance systems facilitate the transition?" (2010) 34(6) *Marine Policy* 1290–1299.

[22] See e.g. the discussion on ecosystem-based legal mechanisms in A. K. Nilsson and B. Bohman "Legal prerequisites for ecosystem-based management in the Baltic Sea area: The example of eutrophication" (2015) 44, Suppl 3 *AMBIO* 370–380.

4.4 The Regulatory Instruments

4.4.1 General Regulatory Aim and Design

The Helsinki Convention is an international agreement of a rather traditional structure. The main obligation of the Helsinki Convention states, *inter alia*, that: "The Contracting Parties shall […] take all appropriate legislative, administrative or other relevant measures to prevent and eliminate pollution in order to promote the ecological restoration of the Baltic Sea Area..."[23] What this implies in regard to eutrophication is elaborated in Annexes and Recommendations attached to the Convention.[24] The regulatory structures of the BSAP and the MSFD are on the other hand based in the goal of "good environmental status".[25] The main requirement in these instruments is for the states to adopt national implementation plans and programs of measures with the aim to achieve good environmental status.[26] The overall goal of the WFD is similarly "good ecological status", although the structure, design and geographical scope of the WFD are differing.[27]

Both the BSAP and the MSFD has a number of focus areas, eutrophication is one such area that these instruments apply to.[28] Both the BSAP and the MSFD translate good environmental status into objectives with descriptors and indicators that characterizes what this means more specifically in each focus area. The goal of good environmental status for eutrophication in the MSFD means that: "Human-induced

---

[23] The 1992 Helsinki Convention, art. 3(1).

[24] Particularly Annex II and III includes guidelines and measures regarding eutrophic pollution from land-based sources, as are HELCOM Recommendations 28E/4, 28E/5, 28E/6, 28E/7.

[25] The MSFD art. 1(1) states that member states shall take the necessary measures to achieve or maintain good environmental status in the marine environment by the year 2020 at the latest. The BSAP is founded upon the HELCOM Vision, which as the aim of *inter alia* Good Environmental Status. This is recalled both in the preamble of the BSAP, as well as in the Eutrophication Segment.

[26] The HELCOM state parties agrees "…to develop and to submit for HELCOM's assessment national programmes by 2010 with a view to evaluate the effectiveness of the programmes at a HELCOM Ministerial Meeting in 2013 and whether additional measures are needed." The BSAP, Eutrophication Segment at p. 9.; The MSFD, art. 1(2) states that marine strategies shall be developed and implemented in order to achieve to over all purpose. The development of marine strategies shall be done in accordance with art. 5.

[27] The WFD, defines as its purpose the protection of inland surface waters, transitional waters, coastal waters and ground-waters, according to art. 1. In order to achieve this ecological objectives are set out in art. 4. A general requirement for ecological protection, and a general minimum chemical standard, was introduced to cover all surface waters in relation to the ecological objectives. Two elements of "good ecological status" and "good chemical status"were then defined. These are elaborated in Annex V of the Water Framework Proposal, in terms of the quality of the biological community, the hydrological characteristics and the chemical characteristics.

[28] The main focus areas of the MSFD are found in Annex I, where Qualitative descriptors for determining good environmental status are found. The BSAP on the other hand is built on four different segments that each has its specific focus area based on the four most crucial problems of the Baltic Sea area, where eutrophication is the first one.

eutrophication is minimised..."[29] The overall goal of HELCOM and the Baltic Sea Action Plan in regard to eutrophication is "...to have a Baltic Sea unaffected by eutrophication."[30] Although the WFD is formulated and structured differently, it still seems to be related and do not differ much in interpretation. The general definitions of ecological status states that high status equals: "...no, or only very minor, anthropogenic alterations to the values of the physico-chemical and hydromorphological quality elements..."[31] It is significant however that the geographical scope of the WFD entails that the WFD does not apply directly to marine waters and thus not to the common areas where the MSFD, the BSAP and the Helsinki Convention applies.[32] Although pollution reductions in coastal areas have important effect also on marine waters, it is not the main purpose or aim of the WFD regulation. The WFD is however to be accounted for in the strategies and measures of implementation of the BSAP and the MSFD. Among its purposes is also to contribute to the achievement of the objectives of relevant international agreements, including those that aim to prevent and eliminate pollution of the marine environment such as the Helsinki Convention.[33] However, this perspective is not included in the requirements on measures to any large extent.

4.4.2   Legal Status and Requirements

The Helsinki Convention is a binding agreement of international law but its provisions are vague. This vagueness is partly due to the general problem of international law connected to the negotiating and agreeing on requirements.[34] The Helsinki Convention provisions should however be read together with its Annexes. The Annexes further define the provisions on pollution and different measures that could be appropriate. This regulatory design is quite common within international law. It could be regarded as a flexible design in that it leaves a large space of discretion to the parties in deciding how to implement the Convention. Both regarding what measures to take and not least in the assessment of what is appropriate relevant measures. In this respect there is no significant difference to the BSAP and the MSFD, although they as a contrast are designed to be flexible and have a more defined basis

---

[29] The MSFD, Annex I, *Qualitative descriptors for determining good environmental status*, para. 5.

[30] The BSAP, Eutrophication Segment, at p. 7.

[31] MSFD Annex V, 1.2, General definitions for ecological status, Table 1.2, Categories "General" and "High status".

[32] The WFD is only applicable to what is defined as "...*coastal and internal waters within the national sovereign.*" Scope and definitions are found in WFD art. 2. More precisely the surface waters on the landward side of the baseline in accordance with UNCLOS and 1 NM on the seaward side of this line. See WFD arts. 2(3) and 2(7).

[33] The WFD, art. 1.

[34] See e.g. T. Gehring, note 9 at 2, at 470; see also A. Chayes and A. H. Chayes *The New Sovereignty – Compliance with International Regulatory Agreements* (Harvard University Press, Cambridge, 1995) at 10–13.

in the environmental status through the ecosystem approach. This is further elaborated in the next section.

The MSFD and the WFD are EU Directives, which implies that they are binding instruments to their aimed result and their regulatory aim shall be transposed into law in the EU member states.[35] The fact that the MSFD and the WFD are framework Directives leaves additional space for discretion in the implementation compared to an ordinary Directive.[36] As stated, the MSFD, the WFD and the BSAP requires that the member states and contracting parties adopt strategies, or programs, for how to reach the goal of good environmental status. A certain scheme is set out for implementation of in the MSFD with a number of steps to be taken in order to assess environmental status as a base for measures to be taken.[37] A similar approach is set out in the BSAP.[38] It is important to note that the BSAP includes a number of guiding statements that further elaborates on the main pollution sources. One of the most interesting statements in this regard is the acknowledgment that agriculture is the main source of nutrient inputs to the Baltic Sea.[39] It furthermore entails the adoption of a number of Recommendations with suggested measures, and additional documents that further elaborates on some of the most crucial issues for abating eutrophication, including a document on measures to reduce nutrient losses from agriculture that is the most crucial pollution sources to the Baltic Sea.[40] An additional requirement within the MSFD, which also gives further emphasis to these features within the BSAP, is that implementation is to be pursued trough regional international organizations.[41] As implied, this is a requirement with consequences of unpredictable dimensions in relation to interpretation and implementation of all legal instruments in the Baltic Sea area.

The BSAP has uncertain legal status in contrast to the Helsinki Convention, the MSFD, and the WFD. The BSAP is an instrument, an action plan, with the purpose to operationalize the ecosystem approach within HELCOM and to be the base for regional operationalization of the MSFD.[42] All the HELCOM state parties have given in specified statements to HELCOM on their acceptance and intent to cooperate in line with the agreed contents of the BSAP. This procedure is in line with the Convention provisions regarding reviews and amendments and similar to the

---

[35] The Treaty on the Functioning of the European Union (TFEU), art. 288.

[36] J. H. Jans and H. H. B. Vedder, *European Environmental Law After Lisbon* 4th (Europa Law Publishing, Groningen, 2012) at 143 ff.

[37] The MSFD, art. 5.

[38] See the BSAP, Eutrophication Segment.

[39] See the BSAP, Eutrophication Segment.

[40] The BSAP, Other Documents 86ff.

[41] The MSFD, art. 6(2).

[42] This is emphasized in many documents and reports by HELCOM. E.g. one of the first: HELOM Ecological Objectives for an Ecosystem Approach, HELCOM Stakeholder Conference on the Baltic Sea Action Plan, Helsinki, Finland, 7 March 2007, states that the BSAP is the tool of implementation of an Ecosystem Approach, p 1, (Introduction); Also the Preamble of BSAP refers both the Ecosystem Approach and to coordinating the work with, *inter alia* the proposed EU Marine Strategy Directive.

procedure prescribed for amendments to the Convention.[43] When an Annex is adopted and amended it is stated very clearly in the Convention that it becomes an integral part of the Convention and thereby is binding on the parties.[44] The status of the BSAP is however not formulated in the Convention provisions, despite a formal adoption. The fact that all the parties have expressively accepted the action plan and with that obligations, including reporting requirements, do speak for at least some kind of obligatory features and good intent. Its status could maybe be comparable to a ministerial declaration, i.e. a formal and norm-setting document although non-binding. However, much of the provisions of the BSAP are formulated vaguely. They are rather on the line of "acknowledging" and "recognizing" the different aspects of the problem, than on "agreeing" on forceful actions. This aspect, of course, makes it easier for the parties to accept and confirm the action plan. The most remarkable provisions of the BSAP are the reduction targets that shall represent the levels of pollution reductions necessary for achieving good environmental status. These are however set out as principal, thus clearly not binding, as opposed to other parts of the provisions. This, again, implies that the parties to some extent do regard the provisions as binding, as there is a clear statement that this part of the contents is not.[45]

4.4.3 Integrating the Concept of Ecosystem Approach

The BSAP and the MSFD are, as stated, intentionally flexible with the concept of ecosystem approach as its main focus. The provisions of the BSAP own special significance since they are defining and operationalizing the MSFD. Although the WFD has a similar structure as the MSFD and the BSAP, the WFD does not expressively state an aim to implement an ecosystem approach. Both the BSAP and the MSFD are set out to be operationalized by a number of ecological and environmental assessments, the first step is to establish through such assessments the current environmental status.[46] The point is to assess the level of the pollution reduction necessary in order to reach a good environmental status. In addition, assessments must also be made in order to define what good status signifies in an ecological perspective. Such definitions are set out in terms of ecological indicators and targets. The BSAP provisional reduction targets are also a way to define such targets, based on previous monitoring and environmental assessments. The targets match

---

[43] The Helsinki Convention, art. 30.

[44] The Helsinki Convention, art. 32.

[45] The reduction targets and other provisions or statements in this respect are found in the Baltic Sea Action Plan, Eutrophication Segment, as "maximum allowable nutrient inputs".

[46] The MSFD, art. 5(2)a (i), art. 8, Annex III; and the BSAP, the Eutrophication Segment and Other Documents, 76.

the indicators for good environmental status.[47] The obligation to adopt measures in order to achieve good environmental status is however difficult to assess despite these indicators. None of the instruments – including the Helsinki Convention – actually expresses what measures states are required to take or how compliance can be evaluated. The reduction targets are an important piece of these structures since it is indirectly articulating what must be accomplished in order to reach the goal, especially seen together with the additional documents and Recommendations adopted with the BSAP. These factors also make the issue of whether the BSAP provisions are binding or not of secondary importance. This part of the BSAP provisions must be considered to also in some extent be defining the implementation of the MSFD. They furthermore contribute to the interpretation of what the aim of the Helsinki Convention and the overarching HELCOM Vision must be. The level of the reduction targets implies that the states need to take far-reaching measures in order to achieve the goal, i.e. good environmental status and minimized eutrophication. The question is whether such flexible approaches are sufficiently effective in steering states towards taking strong measures.

4.5 Flexible Legal Approaches

4.5.1 Background to Integrating the Ecosystem Approach

As stated, the MSFD and the BSAP were both adopted partly as a result of the aim to integrate the concept of ecosystem approach, which is reflected in the design of these instruments.[48] This regulatory approach is the overarching general approach chosen in these instrument but is also more specifically, as part of this, the main approach for regulating eutrophication.[49] As stated above in the section on The Ecosystem Approach in this chapter, it is apparent that the concept of ecosystem approach is based on, or influenced by, general environmental governance theories. The concept embraces a perspective on environment and ecosystems as dynamic and intertwined with human activities. It entails a holistic approach to management.

---

[47] It is stated in the BSAP, Eutrophication Segment, that the parties agree on "...the principle of identifying maximum allowable inputs of nutrients in order to reach good environmental status of the Baltic Sea."

[48] The MSFD, art. 1(3) states that the marine strategies shall apply an "ecosystem-based approach", this is also stressed in the preamble, para. (8). Additionally there are no less than three statements in the preamble of the BSAP that implies the importance of an ecosystem approach. The first recalls "...the 2003 Declaration of the First Joint Ministerial Meeting of the Helsinki and OSPAR Commissions to apply and further develop the measures necessary to implement an ecosystem approach to the management of human activities."

[49] The BSAP embraces four strategic goals, reflecting jointly identified major environmental problems in the Baltic Sea, these goals are "Baltic Sea unaffected by eutrophication", "Baltic Sea with life undisturbed by hazardous substances", "Maritime activities carried out in an environmentally friendly way", which lead to a "Favourable conservation status of Baltic Sea biodiversity". The different focus areas of the MSFD are articulated more specifically for each marine region and are thus included as Quality descriptors in Annex I of the Directive.

The principles of the ecosystem approach emphasizes, for example, flexible and adaptive approaches, stakeholder participation, taking consideration to temporal and spatial scales and propose that management should be decentralized, which are features that can be compared to ideas of, *inter alia,* adaptive management and governance.[50]

The incorporation of such an approach into law is connected to an aim of achieving more effective legal structures. The aim is to have legal structures that are adjusted to the prerequisites set by the environment and the ecosystem through the integration of modern environmental governance approaches that build on research on how the ecosystem responds to change as well as to governance measures. There is, however, an important difference between governance measures and law that makes it uncertain whether this approach really does make the legal structures more effective.[51] Governance is a general structure for measures without specific force and strict obligations, which makes it suitable for flexible and adaptive measures that can be adjusted continuously. The legal nature on the other hand is built on other types of requirements and some extent of force to make actors fulfill obligations although international environmental law is also often considered as providing for governance.[52] This authoritative construction is dependent on foreseeability and clear requirements. This is important in order to be able to plan activities accordingly and as a general expression of the rule of law. Clear requirements could be partly connected to the issue of foreseeability but this is also an issue of having clear obligations in order to also assess compliance. Continuous change and adaptive, flexible approaches could be contradictory to these basic legal features.

Generally, and simplistically, the effectiveness of a treaty is set in relation to whether it achieves compliance and it is ineffective if states fail to meet their obligations. Compliance is then seen as a function of the obligation established by a rule and the actual conduct or results of those subject to the obligation. However, this view of effectiveness is based on two conditions. First that it is clear what a rule imposes. Second, that the goal of the rule is measurable.[53] However, when reviewing complex environmental problems, new governance approaches and the

---

[50] See e.g. R. D. Long, A. Charles and R. L. Stephenson "Key principles of marine ecosystem-based management" (2015) 57 *Marine Policy* 53–60; H. Österblom et al., note 21 at 5.

[51] See e.g. A. K. Nilsson and B. Bohman, note 22 at 5. Many scholars have also debated this approach to environmental law, e.g.: J. B. Ruhl, "Thinking of Environmental Law as a Complex Adaptive System: How to clean up the environment by making a mess of Environmental Law" (1997) 34(4) *Houston Law Review* 933ff.; and A. Wiersema "A Train Without Tracks: Rethinking the Place of Law and Goals in Environmental and Natural Resources Law" (2008) 38 *Environmental Law* 1239ff.

[52] E.g.: P. Birnie, A. Boyle and C. Redgwell *International Law & the Environment 3rd Edition* (Oxford University Press, Oxford, 2009), 9ff and 43ff.; C. Jönsson and J. Tallberg "Transnational Actor Participation in International Institutions: Where, Why, and with What Consequences?", in C. Jönsson and J. Tallberg (eds) *Transnational Actors in Global Governance* (Palgrave Macmillan 2010) at 4ff.

[53] D. Bodansky *The Art and Craft of International Environmental Law* (Harvard University Press, Cambridge, 2010) at 253–255.

ecosystem approach, one can conclude that this is not always the case. If the obligations established are unclear, it will be hard to identify the required (or prohibited) conduct, or if measures taken in the given obligations do not immediately show a result because of time-lags or other natural factors, then the consequence will be that it is impossible to categorize the behavior conducted by the states concerned in terms of compliance or non-compliance.

4.5.2 *Regulatory Approaches*

The legal approaches applied in order to integrate an ecosystem approach, as in the MSFD and the BSAP, include adaptive, decentralized, and goal-based regulatory structures.[54] One main factor in these structures is that the environmental status is the center of the regulatory approach, and thus is meant to steer the choice of measures, actions and other expressions of implementation. This is also a reason why the legal structures become flexible and vague with respect to obligations and requirements.[55] Nor is it defined what measures the states are obliged to take beyond the guidelines found in the Annexes, Recommendations and additional Documents issued by HELCOM.

In addition to requirements to establish a programme of measures and national implementation plans, the MSFD and the BSAP set up a basic structure for monitoring. On the basis of the initial assessments – made as part of developing the marine strategies, coordinated monitoring programs for the on-going assessment of the environmental status of marine waters are to be established and implemented by the member states.[56]

The ecological objectives and associated measurable indicators should be regarded as basic assessment tools linking environmental data to management decisions. The bearing factor in this monitoring structure is that the environmental changes can be assessed and also connected to the measures taken to reduce pollution.[57] The aim of good environmental status should be used as a steering goal and a tool for assessing the measures taken or to be implemented. An important obstacle, however, is the fact that assessing development towards a good environmental status is a complicated issue. Due to natural variations and slow time-scales in measuring reductions in diffuse pollution the concrete anthropocentric contributions to eutrophication are hard to assess. It is difficult to connect cause and effect, especially the measures taken in agricultural practices – the main source of nutrient

---

[54] See section on The Ecosystem Approach above in this chapter and R. D. Long, A. Charles and R. L. Stephenson, note 50 above.

[55] See section on The Regulatory Instruments above.

[56] MSFD, art. 11(1).

[57] E.g. HELCOM 2010 Moscow Ministerial Declaration on the implementation of the HELCOM Baltic Sea Action Plan, May 2010, 10.

input in the Baltic Sea,[58] which are very difficult to assess in terms of changes in the status of marine waters.[59]

It is also acknowledged in the BSAP that the goal of good environmental status will not be achieved on time. HELCOM has stated that due to natural processes, such as time-lags and also climate change, the results of good environmental status will not be assessable by 2021. Although, scientific models indicate significance and rapid improvement as soon as the reduction targets are achieved.[60] In order to reach the goal the indicators must be translated into operational targets relating to concrete implementation measures to support their achievement.[61] The principal reduction-targets could thus still be an important tool. They are also important since they constitute a quantifiable level of reductions that need to be accomplished and point to the kind of measures that need to be taken. The assessments of environmental status, however, do not seem to be an effective tool in enforcing such extensive pollution reductions.

The kind of regulatory structures that are connected to the ecosystem approach have often been discussed and debated. Some argue that adaptive and flexible legal structures represent the only legal design that has the potential to meet complex environmental problems.[62] Others argue that they contradict so-called traditional legal foundations and the main principles of law due to the fact that they are not sufficiently foreseeable and thus do not create stability in the legal system, or that they are not sufficiently effective due to their structure.[63] Today such structures are well established, and just as in the Baltic Sea area, often combined with more traditional forms of regulation. The kind of flexibility and uncertainty that is connected to these regulatory structures are seldom seen as worse problems than other environmental law principles such as the Best Available Technique or the Precautionary Principle, which also change continuously and are interpreted differently depending on the facts applied. These principles could thereby also be regarded as adaptive and flexible without concrete definitions or ensuring total stability and foreseeability for the actors concerned.

What raises questions with regards to the ecosystem approach and how it is applied is the fact that such regulatory instruments are hard to interpret and difficult

---

[58] BSAP, Eutrophication segment at 10.

[59] HECOM BSEP no 89, The review of more specific targets to reach the goal set up in the 1988/1998 Ministerial Declarations regarding nutrients (2003), 9–11; HELCOM BSEP no 100, Nutrient Pollution to the Baltic Sea in 2000 (2005), 14–15.

[60] HELCOM 2013 Copenhagen Ministerial Declaration: Taking Further Action to Implement the Baltic Sea Action Plan – Reaching Good Environmental Status for a healthy Baltic Sea, October 2013, at 6.

[61] HELCOM GEAR 2/2012, Document 4/4, October 2012, at 11.

[62] E.g. W.T. Coleman "Legal Barriers to the Restoration of Aquatic Systems" (1998) 23 *Vermont Law Review* 177ff; and A. D. Tarlock "The Nonequilibrium Paradigm in Ecology and the Partial Unraveling of Environmental Law" (1994) 27 *Loyola of Los Angeles Law Review* 1121ff.

[63] M. Jänicke and H. Jörgens "New Approaches to Environmental Governance" in M. Jänicke and K. Jacob (eds) *Environmental Governance in Global Perspective: New Approaches to Ecological Modernisation* (FFU Report 01–2006, Freie Universität Berlin, 2006).

make into measures to enforce. These factors become more complicated in the Baltic Sea region due to the fact that the MSFD and the BSAP are entangled. The regulatory regimes responsible for enforcing these instruments have divergent control bodies and different degrees of competence when acting to enforce compliance. In addition, it is mainly HELCOM that has the scientific competence and data for any environmental assessment and evaluation.[64] It is also HELCOM that established the provisional reduction targets. Still it is through the MSFD and the EU that the possibilities for more strict compliance control are made possible.[65] Due to the intertwined design of these regulatory structures both the interpretation and implementation of the requirements are affected and more complex. It is, for example, hard to imagine that the reduction targets in the BSAP should not be taken into account when interpreting the obligations of the MSFD, which makes these requirements more far-reaching. The measures set out by BSAP and HELCOM are at the same time given more attention than they would have, had they not been accompanied by a more solid legal regime, such as the EU.

4.6   Enforcement and Goal Achievement

4.6.1   Assessing Compliance

The vague and flexible provisions of the regulatory structures in the Baltic Sea region make it relevant to look deeper into the issues of enforcement and compliance. The most common way to assess compliance has traditionally been to pay attention to the extent to which states comply with their commitments.[66] Hence, to assess the extent of compliance with commitments it must first be clarified what the commitments in the agreement are, which may be difficult in a flexible regulatory instrument. It is sometimes a difficult task to assess compliance when states have a large amount of discretion to interpret and implement an international agreement or can do so due to the vagueness connected to the political negotiation process of a binding document.[67]

The flexibility of legal instruments is even greater if they, as in the case of both the BSAP and MSFD, are action or goal-oriented. Environmental complexity makes

---

[64] HELCOM has issued BSEP Publications since the 1980s and the Pollution Load Compilation Reports frequently publishes reports on the environmental developments in the Baltic Sea in relation to estimated discharge reductions and the current environmental status. See HELCOM BSEP 5b, Assessment of the effects of pollution on the natural resources of the Baltic Sea 1980 (1981).

[65] See further in section on Enforcement and Goal Achievement in this chapter.

[66] E.g. R. B. Mitchell "Compliance Theory – Compliance, effectiveness, and behaviour change in international environmental law", in D. Bodansky, J. Brunneé and E. Hey (eds) *The Oxford Handbook of International Environmental Law* (Oxford University Press, Oxford, 2007) 893–921 at 894 f.

[67] See e.g. the elaborations on Negotiating Agreements in D. Bodansky, note 53 at 11, ch. 8, 154–190.

the situation further complicated and blurred. Compliance is additionally dependent on the ability of states to assess and determine what actions or measures are appropriate.[68] Together with follow-up and review by the Helsinki Convention bodies, it is somehow possible to narrow down and define a somewhat commonly agreed picture of what is expected of the states parties. Compliance is then generally considered to have to be conducted within a certain frame. But then again, many factors affect state behavior and steer it towards either further compliance or obstacles. Such factors are, for example, resources, national strategies for implementation, local incentives and sanctions or other forms of enforcement control.

In legal research one important divide in perspectives on compliance can furthermore be identified, which is important when reviewing issues of enforcement. The first perspective reflects a theory stating that strong enforcement mechanisms are needed. It is based on ideas of *rational choice* and includes a presumption that states will generally be intentionally non-compliant if they can.[69] In practice this theory would imply that the legal structures of the Baltic Sea are inviting the states to be non-compliant since they are hard to enforce strictly due to the basic features building the assessments made by the states themselves, instead of stricter general requirements or prohibitions. The opposing theory argues that states have a predisposition for compliance and therefore non-compliance should be met with a softer approach since it often appears non-voluntary. This theory is often referred to as a *managerial approach*.[70] The latter is built on the assumption that non-compliance generally is due to other factors than intent, i.e. factors that are often beyond the immediate influence of states, for example political changes, a lack of resources or even a misinterpretation of the agreement.[71]

4.6.2   Supervision and Control

Any issues of compliance and enforcement are dependent on treaty supervisory bodies and information about measures or general conduct.[72] All the regulatory instruments in the Baltic Sea region require that the parties report on their implementation to HELCOM and the EU Commission. Treaty programmes to monitor and model environmental processes may improve the quality of data on emissions

---

[68] E. A. Kirk, "Noncompliance and the Development of Regimes Addressing Marine Pollution from Land-based Activities" (2008) 39 *Ocean Development & International law* 235–256, at 235–236.

[69] E.g. R. B. Mitchell, note 66 at 13.

[70] The most famous and widely applicable example of this approach is presented in A. Chayes and A. H. Chayes, note 34 at 7.

[71] The theories arguing for strong enforcement is, e.g., presented in S. Barrett *Environment and Statecraft – Strategy for Environmental Treaty Making* (Oxford University Press, New York, 2003); The managerial opposing perspective was introduced by A. Chayes and A. H. Chayes, A., note 34.

[72] P. Birnie, A. Boyle and C. Redgwell, note 52 at 10, 240.

and implementation and make it easier to set regulatory priorities.[73] Many environmental agreements suffer from poor data gathering. But data also provides the qualitative information for treaty supervisory bodies and for ensuring functional compliance control. The requirements to report and compare data on a regular basis can lead to improvements in the quality and quantity of the data reported, which is also partly how this has developed in the Baltic Sea region.[74] HELCOM has had the task of collecting data from the Baltic Sea coastal states since the 1970s and much of this is now the basis for the assessments made in order to define indicators and targets. HELCOM has subsequently also been appointed as the coordinator for monitoring the region. Previously, however, the monitoring that has been performed by HELCOM has shown considerable shortcomings.[75] During recent years, HELCOM has taken steps to improve their strategy for monitoring to create a more efficient reporting system.[76]

HELCOM is provided with only weak authority when it comes to compliance review. It has the task of supervising and producing recommendations but has no competence to act on alleged non-compliance by any state party, whether it concerns the Helsinki Convention or the BSAP.[77] This is a common feature within international law where the competence to enforce and review compliance is formulated differently in each separate treaty by the states concerned. This means that all parties must specifically (and often separately to other provisions) agree to include a compliance review of some kind.[78] The EU is different in this regard. The EU structure consists of a number of independent EU institutions and bodies, including an internal court system.[79] The EU structure is furthermore based on the competence of conferred powers by the member states.[80] As opposed to international law, whether a legal act is binding or subject to compliance review is not specifically decided on

---

[73] D. G. Victor, K. Raustiala and E. B. Skolnikoff, "Introdcution and Overview" in D. G. Victor, K. Raustiala and E. B. Skolnikoff (eds) *The Implementation and Effectiveness of International Environmental Commitments: Theory and Practice* (MIT Press, 1998, Cambridge and London) at 8.

[74] Ibid., at 18–19.

[75] According to the Helsinki Convention, art. 16(1), the parties have an obligation to report regularly both on measures taken and the effectiveness of such measures. Still, generally very few HELCOM reports can be found that discuss to what extent the measures required were actually implemented. See HELCOM Joint Ministerial Meeting held in Bremen, doc 24/2003: The request for the contracting parties to report on their implementation of HELCOM Recommendations was initially made at the Seventh Meeting of the Commission in 1986 (cf. HELCOM 7/14, para. 3.3). Since 1986 the reporting procedures on how the contracting states should submit information on their implementation of HELCOM Recommendations have been revised within the subsidiary bodies.

[76] HELCOM Ministerial Declaration, Moscow May 2010, Supporting Document, The Report: Implementation of the MSFD in the North Sea and the Baltic Sea – Role of the Marine Conventions HELCOM and OSPAR, 10.

[77] See the Helsinki Convention art. 20.

[78] P. Birnie, A. Boyle and C. Redgwell, note 52 at 10, 250ff.

[79] The Court of Justice of the European Union (CJEU).

[80] In accordance with arts. 5(1) and (2), as well as 4(1) Treaty on European Union (TEU).

in relation to each specific legal act adopted. This lies in the supranational form that signifies the EU, as a result of the principle of conferred powers. The EU competence to review and act on non-compliance is applicable to all EU law, including the implementation of the MSFD and the WFD. The Baltic Sea coastal states that are also EU member states will thus have their implementation reviewed by the EU. Given the link to the BSAP this review is also significant for HELCOM.[81] In this way, the EU regime additionally serves as an important function in the Baltic Sea region. The only state that is not obliged by EU law is the Russian Federation.

4.6.3 Compliance Review and Processes

The managerial approach often includes or promotes mechanisms or compliance procedures that have a softer approach than, for example a court proceeding. Their proceedings and decisions are often less strict and considered non-confrontational in comparison with a court procedure and ruling. A number of environmental treaties have introduced formal non-compliance procedures in this form.[82] The most developed environmental agreements all include regular reporting by the parties, reviews of these reports and frequent on-site inspections. Many agreements also include procedures that allow for some enforcement of international commitments.[83] As implied a soft approach that builds on self-reporting has also been developed within HELCOM, with the obligation for the states parties to report data and measures for implementation. This feature has developed even further with the adoption of the BSAP, which requires not only on-going assessments of environmental status and data but also regular reporting and updating on measures. It is still a fact though that HELCOM lacks the competence to act and enforce commitments in any way. However, as stated, there is an obligation for those states that are EU member states to also report to the EU Commission regarding their implementation of the MSFD (and the WFD), and thus more active enforcement can be triggered in this way.[84]

The EU legal regime is often perceived as more effective in accomplishing a higher level of compliance. In a complex setting like the Baltic Sea it could also be seen as complementary to a more managerial approach.[85] A managerial approach could be effective in the case of complex problems since a way of dealing with issues that are hard to solve that uses more negotiation might be helpful in finding

---

[81] See e.g. J. Ebbesson, "Implementing and Enforcing the Baltic Sea Convention Through European Community Law" in Führ, Whal, and Wilmosky (eds) *Umweltrecht und Umweltwissenschaft: Festschrift für Eckard Rehbinder* (Erich Schmidt Verlag, Berlin, 2007).

[82] P. Birnie, A. Boyle and C. Redgwell, note 52 at 10, 245.

[83] D. G. Victor, K. Raustiala and E. B. Skolnikoff, note 73 at 14, 17.

[84] The MSFD, arts. 5(2), 12 and 13(9).

[85] See J. Tallberg "Paths to Compliance: Enforcement, Management, and the European Union" (2002) 56 *International Organization* 609–643.

solutions, if states have the basic intent to comply.[86] An important difference, and one of the reasons why EU law is often regarded as being more effectively enforced, is the fact that the EU Commission can initiate a court proceeding if a state does not comply with its obligations or if it does not report in accordance with the requirements.[87] This feature of effective control and enforcement is probably also a fact that HELCOM took into account when initiating its cooperation with the EU and the adoption of the BSAP.

4.6.4 Judgments

The EU Commission and Court[88] have so far not issued any judgments that articulate any far-reaching assessments of the content of the plans or measures chosen by the states. Nor have they clarified to what extent the implementation of the WFD will represent the goal of the MSFD, or its relation to the reduction targets and the implementation of the BSAP. The few judgments made in this regard discuss the potential effect that nutrient pollution can have in the marine areas, but mainly mentions that there is an obligation in relation to the BSAP. This is an important development but it is not evident, however, that this affects the judgment.[89] It is also to be noted that these cases then primarily concern point sources in specific sectors, and thus they concern activities that can rather easily be appointed by control measures. From the judgments found it can be identified that the WFD (and its entailed area specific directives) is more strictly implemented than the MSFD.[90] Strong enforcement of the WFD is important, as it is part of operationalizing the MSFD and the BSAP. Both the BSAP and the MSFD does, as previously mentioned, state that the

---

[86] Cf. the general approach within managerial compliance vs rational choice builds on different perspectives on the general reason of why states comply with a treaty. One fundamental stepping-stone in the managerial approach is that it assumes that states have a propensity to comply. Even though there are occasions where states have been deliberately and calculated noncompliant, the most common reasons for non-compliance are generally due to more casual reasons. See further in A. Chayes and A. H. Chayes, note 34 at 7, 3–4 and 10–11.

[87] Art. 17(1) EUT, additionally 258 TFEU about the right to take a member state to the EU Court.

[88] The European Court of Justice (ECJ).

[89] Judgment of the Court (Third Chamber) of 6 October 2009, Case C-438/07, (Commission of the European Communities v Kingdom of Sweden); Judgment of the Court (Third Chamber) of 6 October 2009, Case-335/07 (Commission of the European Communities v. Republic of Finland).

[90] The most significant case in this regard is the Weser Case (Judgment of the Court (Grand Chamber) 1 July 2015, a preliminary ruling under art. 267 TFEU from the Bundesverwaltungsgericht (Germany), where the Court settled on the legal status of the environmental objectives in the WFD and ruled that they are legally binding. Any enterprise that would compromise the objectives or non-deterioration principles is not allowed. This is a ruling that will have important effects on the further implementation of the WFD, and thus indirectly also on the BSAP and the MSFD. However, the environmental objective and the non-deterioration principle will still only be assessed in relation to internal and coastal waters since this is the limitation of the WFD's geographical scope, hence the question is whether this will have a sufficient effect on the marine areas.

measures adopted by the WFD River Basin Management Plans should be included in their implementation. This implies that these measures partly fulfill the aim of the BSAP and the MSFD, thus the enforcement of the WFD also plays a part. It is natural that the WFD measures contribute to the achievement of the goals of the MSFD and BSAP since it concerns the same activities, but it is also clear that it is not sufficient.[91] HELCOM has neither made any specific clarifications on the relationship between the BSAP and the MSFD, nor has it expressed what the reduction targets entail in terms of measures. HELCOM has also focused more of its work on implementing the BSAP in the Russian Federation, which also sends the signal that it is leaving implementation in other states to be handled by the EU.[92]

4.6.5 Managerial Process

In contrast to arbitration and court procedures, the modern kind of treaty supervisory bodies supervise compliance in a collective approach as opposed to the bilateral form that comes with court procedures. The collective approach enables individual states to be held accountable for non-compliance in relation to all the other treaty parties, and the review process may also involve NGOs as observers. The purpose of these treaty bodies is to move away from the strict and authoritative interpretation of treaty compliance towards a non-confrontational mechanism that provides for more negotiation and mutual assistance. A compliance procedure that is more focused on compliance or non-compliance as a process and development.[93] Some of the features that are specific for such non-confrontational compliance mechanisms are, for example, that the supervisory body does not need the consent of the respondent state before a process can be initiated. Furthermore, standing, in the traditional formal sense of being directly concerned, is not condition for making a complaint. In most cases, any party to the treaty or the treaty secretariat may have the right to complain and in some cases NGOs and members of the public may also complain, or provide information to the secretariats.[94]

---

[91] For example, the findings in HELCOM 2013 Copenhagen Ministerial Declaration: Taking Further Action to Implement the Baltic Sea Action Plan - Reaching Good Environmental Status for a Healthy Baltic Sea, 3 October 2013.

[92] On the approach taken by the Russian Federation see e.g. D. Nechiporuk, M. Nozhenko and E. Belokurova "Russia – a special actor in Baltic Sea Environmental Governence" in M. Pihlajamäki and N. Tynkkynen (eds) Governing the Blue-Green Baltic Sea - Societal challenges of marine eutrophication prevention (FIIA Report 31, Tampereen Yliopistopaino Oy – Juvenes Print, 2011, Tampere) 44–54.

[93] As suggested also by E. A. Kirk, note 68 at 13.

[94] P. Birnie, A. Boyle and C. Redgwell, note 52 at 10, 246; A. Epiney "The Role of NGOs in the Process of Ensuring Compliance with MEAs" in U. Beyerlin, P-T. Stoll and R. Wofrum (eds) Ensuring Compliance with Multilateral Environmental Agreements – A Dialogue between Practitioners and Academia (Martinus Nijhoff Publishers, 2006 Leiden/Boston), 319–352.

The underlying idea is that the pressures and the scrutiny of other states in such inter-governmental forum may be more effective in securing a higher level of compliance, given that these relations with other states are important to sustain.[95] A non-confrontational compliance procedure could be seen as a way to avoid dispute or alternative dispute resolution in the sense that a binding third-party court procedure would be avoided, and with that a strict binding decision would also be avoided. It is easier to form a non-compliance review in these alternative compliance mechanisms since the more formal and strict dispute settlement is something that most states are very reluctant to do for several reasons. These procedures are to large extent based on the managerial approach to compliance and the assumption that states normally intend to comply once they have ratified a treaty. They operate on the basis that it is better to assist and encourage a state to improve its actions than to penalize failure. The process is intended to reinforce the stability, transparency and legitimacy of the regime as a whole.[96]

In the Baltic Sea region, the adoption of the BSAP has started a continuous reporting and review process in HELCOM that encourages multilateral communication regarding implementation. With regard to submitting the BSAP National Implementation Programs, HELCOM is to review them and if necessary suggest additional measures. Thus, while HELCOM still has no competence to act on non-compliance, many of the features of managerial compliance can be detected in this setting. In addition, while the EU has the potential to refer a matter to a court procedure, the EU compliance regime could be seen as a combination of features. Before referring a matter to the Court, the EU Commission will initially have a far-reaching managerial communication with a member state. While this is bilateral communication, NGOs and members of the public can at least to some extent make the EU Commission aware of relevant facts connected to alleged non-compliance.[97] Hence, it could be concluded that both HELCOM and the EU include features of managerial non-confrontational compliance review, and these features have become more elaborate with the new flexible design of the BSAP and the MSFD. In the Baltic Sea regulatory structure these features could be seen as additional to the possibility of court procedure. Given the result of reviewing these matters in the EU Court, it is questionable whether it is effective in handling these types of regulatory instruments. In this regard, the flexible design and ecosystem focus of these instruments might imply that their implementation gains from approaches that provide for more communication between the parties of the regulatory instruments. A central issue is whether the EU Commission and HELCOM makes assessments as part of the compliance review to deem whether the measures chosen for implementation are sufficient in order to achieve the goal set out. The basis for such assessments is provided in the guidelines and recommendations connected to the BSAP documents. They

---

[95] P. Birnie, A. Boyle and C. Redgwell, note 52 at 239–240, 246.

[96] Ibid., 245–246.

[97] In accordance with art. 259 TFEU for example, a member state can bring claims on another member state for not fulfilling its environmental obligations under the EU Treaties, and thus initiate judicial review by the court.

were discussed on a general basis and elaborated on in the first implementation review meeting, but were not made strict requirements. One reason for this could be the sensitive nature of these measures and guidelines.

The non-binding nature of the BSAP as well as the lack of competence for HELCOM to act, makes it questionable how far HELCOM can or will demand additional measures in its assessment of the national implementation programmes. It is likely that the stronger incentives for implementation, for most of the parties, are the connection to the MSFD and the WFD. However, so far, the more active and binding review made by the EU has not resulted in any strict compliance or further measures. Since the ecosystem approach creates an important element in these regulatory structures a more flexible communicative procedure for enforcement and implementation could be better adjusted to follow the particular aspects of the implementation and measures that relates to the ecosystem. The environmental indicators and targets do not have any definitive targets and they cannot be treated as having such in the implementation. Of further importance is that it seems as if the BSAP has accomplished putting some of the more crucial issues, such as diffuse pollution from agriculture, on the table for discussion. Although no strict measures have been adopted, it is still significant since this has been an issue where, for a long time, it has been hard to make the state parties agree on more forceful measures. In this way, although the regulatory structures might appear as weak and entailing too much of regulatory gap, they might actually be a tool for initiating pathways that eventually will lead further and be more beneficial than any strict measures or requirements.

4.7   Conclusions

The Ecosystem Approach is a legal concept that aims to integrate general environmental governance features into legal structures. This transition is complicated and entails many questions regarding effective regulatory approach and measures. It also entails a stricter focus on the ecosystem's prerequisites and the environmental status of the problem to be abated. This is also true for the integration of the ecosystem approach into the regulatory structures in the Baltic Sea region. While flexibility was characteristic of the previous regulation regarding the Baltic Sea, the ecosystem approach has added a more direct ecosystem focus to the regulatory approach. This almost comprehensively entails that a more flexible approach is needed in adopting measures and assessing their effectiveness. However, what this focus on the ecosystem also brings into the regulatory structures is an increased vagueness regarding what can be accomplished and how to assess compliance because the environmental or ecological targets and indicators are hard to define, but even harder to evaluate regarding the measures adopted.

This blur increases the burden of the control systems established by these regulatory instruments. It is difficult to assess compliance and how requirements or obligations are to be enforced when they also themselves depend on such vague and continuously changing factors. Whether the goal of good environmental status will be achieved is yet to be seen. Beside the fact that establishing reliable indicators to create a good environmental status is complicated, the environmental prerequisites also entail that natural time-lags will make it impossible to assess the results of the measures taken. In the Baltic Sea region, the issues of assessing compliance and enforcing requirements are even more complicated due to the mix of regulatory regimes and their inter-connectedness that creates further dimensions to how to interpret these instruments and their requirements.

However, while these regulatory designs result in many questions and complication with regard to interpretation, implementation and measurable effects – which makes them questionable in terms of their regulatory effectiveness, they have also created important new incentives for further action with regard to a crucial environmental problem. If the uncertainty entailed with this regulatory design can be bridged and the instruments become well connected, then the intertwined structures and their diverging supervisory bodies could provide important change for the environmental status being focused on. A review of the slow process in moving forward with the on-going implementation and coordination, shows some signs that it is possible to bridge at least some of the uncertainty connected to these approaches.

There is still no clear-cut answer as to how the reduction-targets will affect the level of ambition in the implementation of the MSFD, but important development is taking place. While the EU Court as a mechanism for active enforcement has an important task in relation to the lack of competence shown by HELCOM in acting on non-compliance, it might still be the managerial processes that these new structures have triggered that will entail the most significant results. The continuing work of assessing environmental status, defining indicators, tracking potential environmental change and not least the aim to coordinate all these factors in relation to state implementation also result in slow progress in relation to the issues of diffuse pollution and measures where states have been very reluctant to adopt any strict requirements.

HELCOM lacks the competence to act and thus is left with the only option of a non-confrontational dialogue. The EU Commission on the other had has a general structure of communicating on the issue of alleged non-compliance, which is similar to the structure of a managerial approach. While the last step of this communication is to refer the issue to court, the process of communication is often the main approach taken to make states adjust their behavior in implementing action to reduce marine pollution. The soft approach by HELCOM and the EU Commission has also indirectly brought attention to issues of reducing diffuse pollution related to agriculture. It could be seen as a fault in these structures that this issue, being one of the main issues for abating eutrophication is not more strictly regulated. It is however, due to the fact that this is both a complex and a politically as well as an economically sensitive area to regulate. This means that it is a typical area where managerial compliance is of relevance and, in addition, an area where strict regulatory approaches have failed to accomplish sufficiently strict laws. It might thus be a hopeful and more fruitful path to approach this issue area through the softer method of communicating the legal obligations.

Chapter 5
The Lack of Regulation of Chemical Mixtures and Its Legal Consequences in the Baltic Sea Area

Lena Gipperth and Thomas Backhaus

Abstract The chapter discusses the void in legal protection of the Baltic Sea from the combined effects of multiple chemicals. While individual chemicals are subject to detailed regulation at international and EU-level, combinations sometimes fall outside the scope of any rules. The article analyses to what extent EU laws consider potential combination effects of chemicals and how, which types of chemical exposure is considered and how combination effects of chemicals in the Baltic Sea could be regulated in the future.

Keywords Chemicals · Combination effects · Chemical status · Water framework directive · REACH Regulation · Marine strategy framework directive · Baltic Sea

5.1 Introduction – A World of Chemicals

The environment is exposed to an enormous number of chemical substances. In the EU it is estimated that there are more than 140,000 chemicals[1] and worldwide there are at least 40 new chemicals invented per hour.[2] However, the production and use

---

Our appreciation to colleagues in the research programs NICE (particularly Pontus Nilsson), SPEQS (A systems perspective on environmental quality standards) and the FRAM Centre for Future Chemical Risk Assessment and Management Strategies.

[1] UNEP. 2013. *Global Chemicals Outlook - Towards Sound Management of Chemicals*, 10, available at <http://www.unep.org/hazardoussubstances/Portals/9/Mainstreaming/GCO/The%20 Global%20Chemical%20Outlook_Full%20report_15Feb2013.pdf>

[2] Backhaus, T., Brooks, B. and Kapustka, L. "Chemical Risk Assessment: Pressures, Perceptions and Expectations" (2010) *Integrated Environmental Assessment and Management* 6(3), 323.

L. Gipperth (✉)
Department of Law, University of Gothenburg, Gothenburg, Sweden
e-mail: lena.gipperth@gu.se

T. Backhaus
Department of Biological & Environmental Sciences, University of Gothenburg, Gothenburg, Sweden
e-mail: thomas.backhaus@bioenv.gu.se

of chemicals has two sides. On the one hand, chemicals are crucial for human society as they aid healthcare, food production, the production of consumer goods and support society's general infrastructure. Furthermore, the chemical industry employs approximately 3.1 million people in 83,000 companies in Europe alone.[3] The global revenue generated by the chemical industry was put at US$ 171 billion in 1970 but by 2010 it had grown to US$ 4.12 trillion.[4] On the other hand, the massive production and use of chemicals has substantial negative impacts on the exposed ecosystems life and the activities and services they enable, such as food production and drinking water supply, etc. Many chemical substances end up in the marine environment and today the impact from hazardous substances is regarded as one of the most serious threats to the Baltic Sea ecosystem.

The protection of the oceans from chemical pollution by international law has increased during the last 30 years by agreements such as the 1989 Basel Convention on the Control of Transboundary Movements of Hazardous Wastes and their Disposal, the 1998 Rotterdam Convention on the Prior Informed Consent (PIC) Procedure for Certain Hazardous Chemicals and Pesticides in International Trade and the 2001 Stockholm Convention on Persistent Organic Pollutants (POPs). Also the voluntary scheme GHS (Globally Harmonized System of Classification and Labelling of Chemicals) contribute to the protection by promoting adequate, and reliable information on the hazards of chemicals. There is however no comprehensive international legal framework for chemical pollution, particularly not relating to combination effects. Nationally – but also at EU level – the amount of legislation on the use of chemicals has increased rapidly. Nevertheless, legislation and the assessments of the environmental impacts of these chemicals lag far behind the chemical industry's development of potential new threats to the environment. For instance, the United States Environmental Protection Agency, the world's largest environmental protection agency, only manages to finalize the assessment of less than 20 chemicals *per year*.[5] A further major shortcoming is that the environmental risks of chemicals are assessed and managed individually instead of in combination with other chemicals – the combination effect, which occurs once they enter the environment. The notion that each chemical is the only toxic substance present in an otherwise pristine environment is a dangerous oversimplification of reality, postponing the taking of effective proactive measures. In Sweden, for example, 12 different pesticides are simultaneously found in the average stream[6] and effluents from sewage treatment plants in Stockholm contain more than 137 different organic pol-

---

[3] Kortenkamp, A., Backhaus, T., Faust, M. 2009. *State of the Art Report on Mixture Toxicity*. Report for the Directorate General for the Environment of the EU Commission)

[4] UNEP, note 1 at 11.

[5] Gray, G. M. and Cohen, J. T. "Rethink chemical risk assessments" (2012) 489(7414) *Nature* 27–28.

[6] Backhaus, T., Faust, M. "Predictive environmental risk assessment of chemical mixtures: a conceptual framework" (2012) 46(5) *Environmental Science & Technology* 2564–2573.

lutants.[7] Identical patterns emerge from chemical monitoring studies worldwide.[8] Neglecting combinations of toxic chemicals is particularly problematic because scientific evidence shows that the toxicity of a mixture may exceed the toxicity of each individual compound and that small, individually non-toxic concentrations might when combined create a chemical cocktail that has severe toxicity.[9]

The aim of this article is to provide a general understanding of the void where there should be legal protection for the coastal waters of the Baltic Sea from the combined effects of multiple chemicals. The focus will be on the Baltic Sea Convention and a number of specific legal acts of the EU. The questions to be discussed are:

1. To what degree is the potential combinations effects of chemicals considered and how?
2. Which types of chemical exposure and their effects in the Baltic Sea are regulated?
3. How can the combination effects of chemicals in the Baltic Sea be regulated in the future?

After this introduction there is a short presentation of the background to the problem of chemical mixtures in the Baltic Sea and the important concepts related to this issue. Thereafter, a presentation and an analysis of the current policies and legislation related to the combination effects of chemicals in the Baltic Sea are made. Finally, general conclusions on the scope and quality of the formal legal protection of the Baltic Sea against negative effects of chemical mixtures are presented.

---

[7] Backhaus, T. *An additional assessment factor (MAF) – a suitable approach for improving the regulatory risk assessment of chemical mixtures?* (2015) Report for the Swedish Chemicals Agency, Report May (2015).

[8] Backhaus, T., Faust, M., Kortenkamp, A. "Cumulative Risk Assessment: a European perspective on the state of the art and the necessary next steps forward." (2013) 9(4) *Integrated Environmental Assessment and Management* 547–548. Backhaus, T., Faust, M. *Hazard and Risk Assessment of Chemical Mixtures under REACH: State of the Art, Gaps and Options for Improvement* (2010) PM for the Swedish Chemicals Agency, PM 3/2010.

[9] Fisher, B., Polasky, S., Sterner, T. "Conservation and Human Welfare: Economic Analysis of Ecosystem Services" (2011) 48(2) *Environmental and Resource Economics* 151–159. See also Backhaus, T., Snape, J., Lazorchak, J. "The impact of chemical pollution on biodiversity and ecosystem services: the need for an improved understanding" (2012) 8(4) *Integrated Environmental Assessment and Management* 575–576, and Diamond, M., de Wit, C.A., Molander, S., Scheringer, M., Backhaus, T., Arvidsson, R., Bergman, Å., Hauschild, M., Holoubek, I., Lohmann, R., Persson, L., Suzuki, N., Vighi, M., Zetzsch, C. Exploring the planetary boundary for chemical pollution (2015) 78 *Environment International* 8–15.)

5.2   The Baltic Sea – One of the Most Polluted Sea Area in the World

The Baltic Sea is the second largest brackish water system in the world. Today its eight catchment areas are inhabited by about 85 million people. The Baltic Sea is often described as one of the most polluted sea areas in the world. In a study where the ecological status of twelve estuaries and coastal areas across the world were compared, the Baltic Sea, the Adriatic Sea and Wadden Sea were found to be the most degraded and most affected by the impact of humans.[10] In an assessment of hazardous substances in the Baltic Sea from 2010 it is stated that almost all open-sea areas of the Baltic Sea were classified as "disturbed by hazardous substances" and were granted a status classification of "moderate", "poor" or "bad".[11] Only six of 104 coastal assessment units were classified as "areas not disturbed by hazardous substances" and received a status classification of good or high.[12]

The Baltic Sea also has one of the longest histories of contamination in the world, being exposed to the extensive use of chemicals from the very beginning of the industrialization of the region in the late nineteenth century. The anthropogenic impact on the Baltic Sea ecosystem does not only originate from hazardous substances entering the Baltic from hotspots, it is also the result of minor industrial sources, agriculture with its use of pesticides, pharmaceuticals and fertilizers, households with their extensive use of consumer products, plus the effects from sludge, dump sites and waste disposal and landfill sites. Simultaneously, pressure on the environment comes from stressors like eutrophication, overfishing and intense shipping.

The Baltic Sea is particularly vulnerable to the anthropogenic impact due to several natural factors like low salinity, isolation from other sea areas and low biodiversity, making the ecosystem functions less resilient. The dominant wind directions make the Baltic a recipient for air pollutants from the whole of the European continent, adding to emissions from local sources in the Baltic Sea region. The photodeg-

---

[10]Lotze H.K., Lenihan H.S., Bourque B.J., Bradbury R.H., Cooke R.G., Kay M.C., Kidwell S.M., Kirby M.X., Peterson C.H. & Jackson J.B.C. "Depletion, degradation, and recovery potential of estuaries and coastal seas" (2006) 312(5781) *Science* 1806–1809.

[11]HELCOM, 2010. Hazardous substances in the Baltic Sea. An integrated thematic assessment of hazardous substances in the Baltic Sea. *Baltic Sea Environment Proceedings* No. 120B, north-western Kattegat was the only exception which was classified with a "good" status.

[12]Coastal areas that received the highest status classifications were found at wave-exposed sites such as the Åland Archipelago and the coastal waters off Kaliningrad. Assessment units with the poorest status were either located near big cities or ports (Tallinn, Klaipeda) or in estuarine areas (Ruotsinpyhtää in the Gulf of Finland), Kvädöfjörden in the Western Gotland Basin) or coastal sites (the Kiel Bay area). All common groups of hazardous substances – PCBs, dioxins, heavy metals, organometals, alkylphenols, phthalates, brominated substances, polycyclic aromatic hydrocarbons (PAHs), DDTs and chlorinated pesticides as well as the radionuclide cesium-137 – were found among the substances with the highest concentrations in relation to target levels. See HELCOM, 2010. Hazardous substances in the Baltic Sea An integrated thematic assessment of hazardous substances in the Baltic Sea. *Baltic Sea Environment Proceedings* No. 120B.

radation of hydrophobic organic contaminants is also influenced by the low temperature as well as eutrophication, which cause hypoxic and anoxic conditions below the halocline.[13]

The Helsinki Convention on the Protection of the Marine Environment of the Baltic Sea Area was signed in 1974 and is mainly aimed at preventing and eliminating pollution caused by hazardous chemicals. The Convention has been the driver for many initiatives taken to reduce the discharge of pollutants in the Baltic Sea. Between 1993 and 2009 the HELCOM list of the hot spots of the main polluters in the region was reduced from 163 to 50. Over time, and particularly after the new convention was signed in 1992, the work within HELCOM has broadened to include eutrophication, biodiversity and the environmental safety of maritime activities. The HELCOM Baltic Sea Action Plan adopted at ministerial level in 2007, makes it clear that hazardous substances remain a focal area of HELCOM.

Also in the EU, the control of hazardous substances has been a priority topic within environmental policy. In the late 1990s, concern arose over a deficient EU chemicals policy and reform was initiated, leading to the establishment of two new European agencies responsible for the risk assessment of chemicals, the European Chemicals Agency (ECHA) in Helsinki and the European Food Safety Authority in Parma (EFSA). Several new major European Regulations entered into force, like Regulation 1907/2006/EC concerning the Registration, Evaluation, Authorisation and Restriction of Chemicals (REACH), Regulation 1272/2008/EC on Classification, Labelling and Packaging (CLP Regulation), Regulation (EC) No 1107/2009 concerning the placing of plant protection products on the market and the Biocidal Products Regulation 528/2012 (BPR). Through directives related to environmental status, (like the Water Framework Directive 200/60/EC and the Marine Strategy Framework Directive 2008/56/EC), to production (such as the Directive 2010/75/EU on industrial emissions (integrated pollution prevention and control) as well as waste (like Directive 2008/98/EC on waste) the EU has developed a legal framework covering the lifecycle of chemicals, but without addressing the full scale of the toxicity of chemical mixtures. Before presenting this framework and analysing its gaps, the concepts of mixtures and the effects of chemicals when combined will be briefly presented.

5.3   A Conceptual Introduction

The use of terms in relation to combination effects is not consistent, when comparing different situations, legislation and legal systems. From a legal perspective, it is evidently crucial that the concepts – cocktails, mixtures and combination effects – do not create confusion and are interpreted and applied in the way legislators intended.

---

[13] Magnusson, K. and Norén, K. *The BaltSens project: The sensitivity of the Baltic Sea ecosystems to hazardous compounds* (2012) KEMI.

Effects from mixtures can, depending on exposure conditions, exposed organisms and chemicals involved, act independently or have concentration-additive, synergistic (more than additive) or antagonistic (less than additive) effects.[14] In particular, the following situations can be distinguished[15]:

1. **Similar action**: For similar acting substances, the effects can be calculated by adding their toxicity-adjusted doses/concentrations in a mixture (dose/concentration addition). *Example:* A person drinks beer and wine in a bar. The same type of toxicant (alcohol) through the same route or pathway intoxicates the person. The total effect can be described as dose additive, in this case, alcohol + alcohol.
2. **Dissimilar action:** For independently acting substances, the total effect can be estimated from the responses from the individual components (response addition, also termed independent action). *Example:* A person drinks beer and smokes a cigarette. The person is intoxicated through different routes or pathways. The total effect can be described as independent = alcohol + nicotine.
3. **Interactions**: The effect of several substances that are stronger (synergistic) or weaker (antagonistic) than expected.[16]*Example:* A person drinks a beer but feels a light headache and takes a painkiller that decreases their liver function, making the person more intoxicated than expected after just one beer. The effect can be described as an interactive effect, in this case a synergistic effect.

Regarding coastal waters, polluting substances can originate from a range of sources, for example, wastewater, agriculture, industry, traffic and the disposal of pharmaceuticals or other products. The combined discharges produce the total pollution levels in the recipient. This situation can be described as multiple substances from multiple sources or *combined exposure*.[17] Multiple exposures present a regulatory dilemma if several sources lead to the same recipient, such as the Baltic Sea. Each one contributing within acceptable limits, but together exceeding what the recipient's ecosystem can tolerate. This situation raises many questions about burden sharing and fairness, particularly when considering whether the exposure occurs simultaneously or in sequence.

---

[14] Backhaus, T., Faust, M. and Blanck, H. *Hazard and Risk Assessment of Chemical Mixtures under REACH. State of the Art, Gaps and Options for Improvement* (2010) PM 3/10.

[15] SWECO VIAK *Kombinationseffekter av. föroreningar* (2008) Naturvårdsverket 19.

[16] SCHER, SCCS, SCENIHR. *Opinion on the Toxicity and Assessment of Chemical Mixtures* (2012) 9, and SWECO VIAK, note 15 at 9.

[17] SCHER, SCCS, SCENIHR, note 16 at 43. In the US this situation is called *cumulative exposure*. In Europe, *cumulative exposure* denotes the release of only one substance from one or several sources through one or several routes. In the US the term for this situation is *aggregated exposure*. Hazard and risk assessment of chemicals under REACH, 47.

5.4   Current Policies and Legislation Related to the Combination Effects of Chemicals

5.4.1   International Regulation in Relation to the Baltic Sea

Although oceans are the recipients of pollution from many chemicals there is no international legislation directly addressing the issue of the combination effects of chemicals in oceans. There are several international conventions and protocols regulating single substances or groups of substances. The Stockholm Convention on Persistent Organic Pollutants (POPs) came into force in 2004 with the objective of protecting human health and the environment from persistent organic pollutants. Contracting parties are required to take measures to eliminate or reduce the release of POPs into the environment. The International Convention on the Control of Harmful Anti-fouling Systems on Ships (AFS convention) was adopted in 2001 as a response to knowledge about the severely negative effects on the marine environment of using tributyltin (TBT) in anti-fouling paints. The AFS convention is so far only applicable to TBT but might be extended by the parties to other toxic anti-fouling substances. The UNECE Convention on Long-range Transboundary Air Pollution entered into force in 1983. This convention sets a framework for regulating an extended number of air polluting individual substances that contribute to acid rain, like ground-level ozone, persistent organic pollutants, heavy metals and particulate matter.

The International Convention for the Prevention of Marine Pollution from Ships (MARPOL) regulates pollution from ships and the carriage of chemicals by ships but does not address chemical mixtures in the oceans that shipping might contribute to.

From a Baltic Sea perspective, the most important international regulation in relation to chemicals is the Helsinki Convention from 1992 (Convention on the Protection of the Marine Environment of the Baltic Sea Area), replacing a 1974 convention with the same name. The convention requires the contracting parties to take all appropriate measures to prevent and eliminate pollution caused by harmful substances from all sources in order to promote the ecological restoration of the Baltic Sea area and the preservation of its ecological balance. This requirement also includes the risks of chemical mixtures but the specific bans and restrictions taken under the convention have so far only been related to a single hazardous substance or groups of them, like DDT and its derivatives DDE and DDD for all final uses except drugs and PCBs and polychlorinated terphenyls (PCTs). HELCOM has issued a great number of recommendations relating to single hazardous substances. In 1998, one recommendation listed 280 chemicals as potential substances of concern to be considered and 42 were then prioritised for action, including pesticides, metals and other industrial substances, e.g., nonylphenol.[18]

---

[18] HELCOM. **Recommendation 19/5 (1998). HELCOM objective with regard to hazardous substances**.

In order to ensure the Baltic Sea has a good environmental status by 2021, the Baltic Sea Action Plan – BSAP – (HELCOM 2007a) in which one of the strategic goals is a Baltic Sea undisturbed by hazardous substances, was decided on in 2007. The aim of the Baltic Sea Action Plan is that concentrations of hazardous substances should be close to natural levels, all fish safe to eat, wildlife healthy and radioactivity at pre-Chernobyl levels. Even if not explicitly mentioned, it is necessary to consider and include combination effects when trying to achieve these objectives. There are, however, no further measures directly addressing the issue of chemical mixtures in the BSAP. Although not all parties to the Helsinki Convention are members of the EU, the Marine Strategy Directive (MSFD) has become an important instrument for implementing the BSAP. HELCOM has a role as a coordinating platform for the regional implementation of the MSFD in the Baltic Sea and works towards joint indicators and coordinated monitoring in order to provide coherent assessments of the status of the Baltic Sea and the pressures and impacts affecting its status. Strong enforcement mechanisms within the EU promote not only the implementation of the MSFD but indirectly the BSAP as well.

It can be concluded that there are no international regulations directly protecting the Baltic Sea from the negative effects of chemical mixtures. Although it does not directly address the status of the Baltic Sea, it is worth mentioning an example of an international agreement that takes into account the combination effects of chemicals – the UN Framework Convention on Climate Change (UNFCCC) and its protocol. The aim with the UNFCCC is to stabilise greenhouse gas concentrations "at a level that would prevent dangerous anthropogenic interference with the climate system". The requirements of the parties are related to the total impact of these gases. There would thus be a possibility of managing toxic chemicals and regulating the total toxicity of chemicals likewise.

5.4.2   EU Legislation Relating to the Combination Effects of Chemicals in the Sea

The problem of combination effects of chemicals has been acknowledged by the EU Council of Environment Ministers in 2009 that "further action in the field of chemicals policy, research and assessment methods to address combination effects of chemicals is required" asking the EU Commission to "review the existing research base".[19] After being presented by a state of the art report on mixture toxicity in

---

[19] Council of the European Union *Combination effects of chemicals – Council conclusions* (2009) 17820/09, available at http://register.consilium.europa.eu/doc/srv?l=EN&f=ST%2017820%20 2009%20INIT

2010,[20] and a scientific opinion on the issue by its scientific committees,[21] the EU Commission adopted a communication,[22] in which was concluded that: *"Current EU legislation does not provide for a comprehensive and integrated assessment of cumulative effects of different chemicals taking into account different routes of exposure."*[23]

The Commission stated that the strict limits for the amounts of particular chemicals allowed in food, water, air and manufactured products, rarely take into account the potentially toxic effects of these chemicals in combination. The Commission further committed itself to improving the understanding of the mixtures and to ensure that EU legislation delivers consistent risk assessments for priority mixtures.[24]

It is thus clear that there is a general void in the legal protection given by EU regarding pollution from chemical mixtures in the Baltic Sea. Nonetheless, there are examples of how EU legislation has developed to handle this challenge. Below follows a short summary of how the challenges of the combination effect of chemicals in marine waters, are handled in some of the key EU chemical regulations. The legislation demonstrates several types of legal approaches, regulating products, chemical substances, environmental status (water) and different types of actors. As this legislation has different aims, the legal basis in the EU treaties for deciding on it differs.

5.4.2.1 Directive 2000/60/EC Establishing a Framework for Community Action in the Field of Water Policy – (WFD)

In order to develop the comprehensive management and protection of water resources, the EU decided on the Water Framework Directive (WFD) in 2000. This directive applies to all surface water, ground water and coastal waters (1 nm out from the baseline) within the EU. Water should, according to the WFD, be managed by river basins and not administrative or political boundaries. The aim is to achieve no further degradation of water status and at least a good water status – both in terms of chemical and ecological status – in all water bodies by 2027. Member states are required to take all appropriate measures to achieve the objectives and to decide on river basin management plans, including programmes of measures.

Good chemical status is defined as compliance with the EU environmental quality standards (EQS), both new and old, as well as the national EQS decided by the member states. In relation to the WFD, EQS has been established for 45 prioritized

---

[20] Kortenkamp, A., Backhaus, T. and Faust, M. *State of the Art Report on Mixture Toxicity* (2010) available at http://ec.europa.eu/environment/chemicals/effects/pdf/report_mixture_toxicity.pdf

[21] SCHER, SCCS, SCENIHR, note 16.

[22] European Commission *Communication from the Commission to the Council. The combination effects of chemicals. Chemical mixtures* COM(2012) 252 final.

[23] Ibid., s. 5.1, (1)

[24] Ibid., s. 5.1, (2).

chemical pollutants by the EQS Directive 2008/105/EG.[25] The WFD offers the explicit option to define EQS for groups of substances and this has been done in some cases, e.g. an EQS for 29 dioxins and dioxin-like compounds in biota defined in terms of the sum of toxic equivalents (TEQs).[26] Most of the EU Priority Substances do not have quality standards for sediments or biota, as quality standards have mainly been defined for the water phase. Assessments, however, show high concentrations of PCB, tributyltin (TBT), mercury and cadmium in fish, mussels and sediment all over the Baltic Sea area. The EQS for water in Sweden has recently been supplemented with criteria for some substances in sediments and biota.[27]

Good ecological status is defined by the quality of the biological community, the hydrological characteristics and the chemical characteristics. Due to ecological variability, the biological criteria are to be decided separately for each water body. The WFD provide a complicated process to set parameters for water bodies and how to combine the single status assessment with an overall assessment. The principle of "one out all out" clarifies that all parameters must be met and that the parameter with the lowest status determines the overall status.

There are no clear combination considerations expressed in the WFD itself and the default approach is to handle each chemical substance by itself. There have been several proposals for setting EQS for priority mixtures, although the legal base under the current WFD for such EQS is not clear.[28] However, the determination of good ecological status (including some chemical criteria) provides a way to consider a complex variety of different criteria that affect ecological status. Although the chemical status criteria do not at all reflect the same considerations for complexity in water, it still points to the potential to also deal with a toxicity standard in the future, especially when considering the combination effect of chemicals occurring in water, sediments or biota.[29]

5.4.2.2 Directive 2008/56/EC Establishing a Framework for Community Action in the Field of Marine Environmental Policy – (MSFD)

The Marine Strategy Framework Directive (MSFD) of 2008 is a WFD media-oriented act, which focuses on the status of the ecosystem. The aim is to protect the marine environment by achieving good environmental status (GES) by 2020. To

---

[25] The directive was updated in 2013 with 12 new substances (2013/39/EU).

[26] There are also EQS referring to a sum of concentrations of selected congeners or isomers of a parent compound: six PBDEs, four cyclodiene pesticides and four DDT isomers.

[27] Regulation from the Agency for marine and water management (*Havs- och vattenmyndighetens föreskrifter* - HVMF 2013:19).

[28] Carvalho, R.N., Arukwe, A., Ait-Aissa, S., Bado-Nilles, A., Balzamo, S., Baun, A., et al. "Mixtures of chemical pollutants at European legislation safety concentrations: how safe are they?" (2014) 141 *Toxicological Sciences* 218–233.

[29] This option has also been discussed, e.g. in European Commission *Guidance Document No. 27. Technical Guidance For Deriving Environmental Quality Standards* (2010).

determine what GES is, 11 descriptors shall be used by the member states. Descriptor 8 is the most relevant for chemical pollution and concerns concentrations of contaminants at levels not giving rise to pollution effects. The guidance document on criteria and methodological standards points out the close relation to the WFD and its list of prioritised substances in Annex X and the EQS directive and EQSs derived in accordance with Annex V in the WFD on the member state level. Other relevant EU legislation is also to be taken into account and, in that regard, the guidance document specifically mentions REACH.[30]

In the initial assessment of the Marine environment the member states must take cumulative and synergistic effects into account according to art. 8. This consideration goes, however, beyond the assessment of combination effects of chemicals, and aims at the assessment of multiple stressors in general, e.g. combinations of underwater noise, pollution and eutrophication that together might cause a cumulative or synergetic effect on the ecosystem. How effects from several pressures from the same descriptor should be handled is unclear, but the work of determining relevant methods is ongoing. It is interesting to note the much clearer way MSFD expresses concerns on combination effects when compared to WFD.

HELCOM has compiled a set of criteria for determining GES in the Baltic Sea.[31] This is called the CORESET in which lists of substances that are of most concern for the marine environment are presented. Combination effects seem to be regarded to some extent, for example for polyaromatic hydrocarbons (PAHs).[32]

5.4.2.3   Regulation (EC) No 1907/2006 Concerning the Registration, Evaluation, Authorisation and Restriction of Chemicals – (REACH)

REACH is the most important legislation to control the production and use of industrial chemicals. This regulation entered into force in 2007 but will not be fully operational until 2018. The legal base of REACH is art. 114 in the TFEU and points at its value as an instrument to harmonise the market and it thus has a different background to the WFD and the MSFD. The purpose of REACH is, *"to ensure a high level of protection of human health and the environment, including the promotion of alternative methods for assessment of hazards of substances, as well as the free*

---

[30] See *Commission decision of 1 September 2010 on criteria and methodological standards on the good environmental status of marine waters* (2010/477/EC), in the Annex part B, descriptor 8 and the European Commission *Commission staff working paper. Relationship between the initial assessment of marine waters and the criteria for good environmental status* SEC (2011) 1255 final, 55.

[31] HELCOM, 2013. HELCOM core indicators: Final report of the HELCOM CORESET project. *Baltic Sea Environment Proceedings* No. 136 I. Also OSPAR has produced a guidance document specifically for descriptor number 8. OSPAR - *MSFD Advice document on Good environmental status - descriptor 8: Contaminants, a living document – Version of 2 March 2012)*. Approaches to determining good environmental status, the setting of environmental targets and selecting indicators for Marine Strategy Framework Directive descriptor 8.

[32] Ibid.

circulation of substances on the internal market while enhancing competitiveness and innovation."[33]

REACH applies to the whole lifecycle of many chemical substances, mixtures and articles used today in industrial and consumer use.[34] The company that has responsibilities according to REACH has the burden of proof that it has fulfilled its duties, but receives some guidance from ECHA. REACH does not by itself regulate emissions into the environment but four main mechanisms by which the use and trade of chemicals inside the EU are controlled:

1. *Registration* is mandatory for everyone who produces or imports a substance, mixture or article and puts it on the market in the EU, in amounts that exceeds 1 tonne per year.[35] Providing sufficient information on the subject is a prerequisite for registration, depending on the quantity of the chemical that shall be used per year. If the registered amount is more than 10 tonnes per year, a Chemical Safety Assessment (CSA) must be performed.[36] If the assessed substance is dangerous, the assessment must include exposure estimation and a risk characterisation.[37]
2. *Evaluation* of substances under REACH aims to create a database for further action.
3. *Authorisation* is used to control the use of dangerous substances regardless of the amount produced or imported.[38] Chemicals subject to the mandatory authorisation process are also subject to a demand for substitution if less hazardous alternatives are or become available.[39]
4. *Restriction* of a substance by itself or particular uses of that substance is possible under REACH.[40] The restriction process includes a socio-economic assessment as well as an assessment of the availability of alternative substances.[41]

REACH focuses on substances which are either pure chemicals, or complex mixtures (so-called Multi-Constituent Substances (MCS) or Substances with Unknown or Variable Composition, or of Biological Origin (UVCB's, such as e.g. fermentation products). MCS and UVCB's are in principle evaluated similar to pure chemicals substances. Final articles that contain chemicals mixtures are not interesting from a registration perspective, only the substances they contain.[42] In 2010 there was a change in REACH (Annex II) demanding information regarding certain types

---

[33] Art. 1.1 REACH.

[34] Art. 1.2 REACH.

[35] Arts. 6, 7 and 10 REACH.

[36] Art. 14 REACH.

[37] Art. 14.4 REACH. The exposure assessment includes both the production and use of the substance, as well as the waste stage. For more detail on how an exposure assessment is done and its function on the chemical safety report, see Annex I, section 5 in REACH.

[38] Art. 56, REACH.

[39] Arts. 60.5 and 61.2(b), REACH.

[40] Art. 67, REACH.

[41] Art. 68, REACH.

[42] ECHA *Guidance on registration* (2012) ECHA-12-G-07-EN, 14.

of combinations of toxic effects, to be included in the CSA. However, this require-ment is limited to situations of relevance for human health but not to ecosystems in coastal waters.[43]

REACH does not affect the application of the WFD or other environmental pro-tection legislation from the EU.[44] Nevertheless, the authorisation of a dangerous substance can be reviewed at any time if the status of surface water, set by the WFD, is endangered.[45] Reviewing is an option if the use of a substance affects the fulfil-ment of the objective to achieve good water status and jeopardises the achievement of EQS set out in WFD. This link to the WFD opens up the nature of REACH's specific area of regulation and links it to the ecosystem-based view in the WFD.

5.4.2.4   Regulation (EU) No 528/2012 Concerning the Making Available on the Market and the Use of Biocidal Products – (BPR)

Another piece of product-related EU legislation is the Biocide Product Regulation (BPR). As with REACH, BPR is based on art. 114 TFEU and the precautionary principle and replaced directive 98/8/EC in 2013.[46] The purpose of BPR is to enhance the inner market by harmonising the regulation of market-access and the use of biocide products and goods treated with biocides as well as to create a high level of protection for humans and the environment.

The BPR provides two main mechanisms; the approval of active substances and the approval of biocidal products.[47] In order to approve the active substance and biocidal product, special requirements needs to be met.[48] There is, however, a bal-ance between the negative effects of the biocide itself and the negative effects in society of not using the biocide.[49] If the intended use of the biocide product under-mines the achievement of compliance with environmental objectives in WFD or MSFD, a member state shall not grant the approval of the product.[50]

For *product evaluation* the terms *cumulative* and *synergistic effects* are used. *Cumulative* expresses both *aggregated exposure* (one substance, all routes) and *mix-

---

[43] Cumulative exposure (one substance, all routes) to the same substance is handled in Annex I (section 6) and ECHA *Guidance on Information Requirements and Chemical Safety Assessment. Part E: Risk Characterisation* 2016 part E 4.5, step 5.

[44] Art. 2.4 in REACH.

[45] Art. 61.5 in REACH.

[46] Arts. 1.1 and 97, BPR. Substances that are authorised in accordance with directive 98/8/EC can also be authorised in the new BRP (art. 86).

[47] The listed biocide products that are covered by BPR need to be approved under BPR but are not subject to authorisation under REACH, Annex V of BPR. Products and treated goods covered by other specific legislative acts, such as the plant protection regulation, are outside the application area of the BPR (art. 2). See the preamble para. 11 in BPR.

[48] Arts. 4.1 and 19 of the BPR.

[49] Art. 19.5 in BPR.

[50] Annex VI paragraphs 67 and 69 in BPR. Approval should not be granted if the product does not fulfil the specific criteria of effectiveness, effects, etc. (art. 19 (iv)).

ture toxicity.[51] This means that the combined toxicity of the product shall be assessed with regard to all its components, regardless of their mode of action, for which a specific interim guidance document has been published by ECHA.[52] Exposure to a recipient is considered with regard to the total pressure of the product itself, but not other biocidal products. To approve an active substance, the *cumulative* effects of both the product and other products must be considered.[53] It must also be possible to use the active substance in a biocidal product to fulfil the requirements in art. 19.[54] This generates a wider assessment where exposure to other products can be included and the combined exposure (multiple chemicals from multiple routes) is included.

5.4.2.5 Regulation (EC) No 1107/2009 Concerning the Placing of Plant Protection Products on the Market – (PPR)

The third product-related EU legislation that is relevant, when discussing voids in managing combination effect toxicity in the Baltic Sea region, is the regulation of plant protection products (PPR), operational since mid-2011.[55] Plant protection products are products that have the purpose of protecting plants from pests or other unwanted plants.[56] PPR is based on art. 114 TFEU and is thus primarily a free market based regulation. The purpose of PPR is not only to harmonise the rules on marketing plant protection products and improve agricultural production, but also to ensure humans, animals and the environment a high level of protection.[57] PPR is limited to areas used for outtake of drinking water or levels of prioritized substances, and not surface water areas.

PPR is linked to the WFD. If a target related to an environmental objective under the WFD is endangered, the Commission can review the approval of an active substance and a member state shall review a product's approval.[58] For other environmental objectives, such as the GES of surface water, the review of an approval is not applicable. Instead, alternative legal grounds for a review are possible in art. 4.3 (e)

---

[51] Art. 19, BPR.

[52] ECHA *Transitional Guidance on mixture toxicity assessment for biocidal products in the environment* (2014).

[53] Art. 8.3, BPR.

[54] Art. 4.1, BPR.

[55] This regulation repeals the Council Directives 79/117/EEC and 91/414/EEC. These older directives had few considerations regarding the mixing of several compounds and their potential combination effects. Instead the assessments were based on the properties of individual substances.

[56] In art. 2 the area of application is set. Products that affect the lifecycle or serviceability of plants can also be viewed as a plant protection product. The relation between substances regulated in PPR and REACH is described in art. 56.4 (a) and (b) REACH.

[57] Art. 1.3, PPR.

[58] Arts. 21 and 44.2 PPR. As legal grounds for review, only arts. 4.1(a)(iv) protection from priority hazardous substances, 4.1 (b)(i) protection of groundwater, 7.2 and 7.3 protection of drinking water, in WFD are relevant.

(i) PPR, but the effects on the environment must then be unacceptable. Measures taken under this directive shall be supportive and in harmony with both the provisions in PPR and WFD.[59]

In PPR the specific consideration of combination effects is limited to the assessment of human health effects (art. 4.2a). However, it has been argued that the uniform principles as laid down in Regulation 546/2011/EC implementing PPR require authorization of plant protection products to be based on the "proposed conditions for use" and consequently – given common agricultural practice – to consider the environmental impact of the resulting pesticide mixtures.[60] This might be a reason why detailed guidelines on the assessment of combination effects are provided by EFSA,[61] and the active *substance* assessment in the PPR is including combination-effects between active substances as well as between active substances and other ingredients of the formulated plant protection products.

5.4.2.6 Directive 2010/75/EU on Industrial Emissions (Integrated Pollution Prevention and Control) – (IED)

In addition to EU regulation relating to environmental status (WFD and MSFD) or products (REACH, BPR and PPR), a third group relates to activities. An important example in the context of managing the pollution caused by chemical mixtures, is the 2010 Industrial Emission Directive (IED). It superseded the Integrated Pollution Prevention and Control directive 2008/1/EC (IPPC) and six sectorial directives regulating emissions from industrial and energy producing installations.[62] Industrial activities listed in the directive are required to have and operate in accordance with a permit, granted by an authority in the Member State. Permit conditions including limit values for industrial emissions are set at the level of the best

---

[59] See the preamble, para. 3. Linked to PPR is Directive 2009/128/EC, establishing a framework for Community action to achieve the sustainable use of pesticides. The Directive sets provisions for how to use, store and apply pesticides. It complements the WFD and points out the responsibility of member states to avoid certain substances, prioritised according to the WFD (art. 11 in directive 2009/128/EC).

[60] Frische, T., Matezki, S. & Wogram, J. "Environmental risk assessment of pesticide mixtures under regulation 1107/2009/EC: a regulatory review by the German Federal Environment Agency (UBA)" (2014) *Journal für Verbraucherschutz und Lebensmittelsicherheit.*

[61] EFSA "Risk assessment for birds and mammals" (2009) 7(12) *EFSA Journal* 1438 (doi: doi:https://doi.org/10.2903/j.efsa.2009.1438). EFSA "Scientific Opinion: Guidance on tiered risk assessment for plant protection products for aquatic organisms in edge-of-field surface waters" (2013) 11(7) *EFSA Journal* 3290 (doi: https://doi.org/10.2903/j.efsa.2013.3290). EFSA "Guidance Document on the risk assessment of plant protection products on bees (Apis mellifera, Bombus spp. and solitary bees)" (2013) 11(7) EFSA Journal 3295 (doi: https://doi.org/10.2903/j. efsa.2013.3295). EFSA "Scientific Opinion addressing the state of the science on risk assessment of plant protection products for non-target terrestrial plants" (2014) 12(7) EFSA Journal 3800 (doi: https://doi.org/10.2903/j.efsa.2014.3800)

[62] See Swedish Government Bill 2012/13:35 p. 20.

available technique (BAT).[63] The level of BAT is what must be reached, taking into account geographical and local environmental conditions.[64]

To support the understanding of BAT, conclusions on the best available technique for a specific industrial sector are compiled and published. The authorities granting permits in the member states must consider these documents when deciding on permit conditions.[65] BAT conclusions are regularly updated.[66]

The IED builds on the integrated approach, acknowledging the whole environmental performance of an industrial plant. EQS set according to, say the WFD, are taken into account when deciding on permit conditions. However, stricter conditions than BAT should be decided on if so required for an EQS.[67]

IED lacks combination considerations in the directive itself. Nevertheless, the determination of BAT provides a branch specific description of the current standards with regard to both the technological development and environmental concerns. This increases the flexibility to include combination effect considerations in each BAT conclusion, if appropriate and needed.

5.5   Analysis of the Existing Legal Instruments for Handling the Combination effects of Chemicals in the Baltic Sea

Together, international law and the six EU acts presented above, cover an extensive area for the control of chemical pollution in the Baltic Sea. These acts represent different legal approaches and consider combination effect in different ways:

- Helsinki Convention: general requirements to prevent and eliminate pollution caused by harmful substances, by BSAPs objective linked to descriptor 8 of MSFD,
- Water Framework Directive: with regard to deriving EQSs for specific pollutants according to the ecological status required
- Marine Strategy Framework Directive: in the risk-based approach for setting criteria for descriptor 8 and for combination effects between different types of pressures on the marine environment as described by different descriptors,
- REACH: in the Chemical Safety Assessment for MCS and UVCB substances required in registration and authorisation situations and the Safety Data Sheet,
- Biocidal Products Regulation: in the assessment connected to the approval of active substances and biocide products,

---

[63] Art. 9.4, IPPC.

[64] See art. 9.4, IPPC.

[65] Art. 14(3) IED. See the European Commission Joint Research Centre website http://eippcb.jrc. ec.europa.eu/reference/. Under the IPPC directive, the BAT conclusions were called BAT reference documents (BREFs) and had a weaker legal value in the national permit processes.

[66] Art. 13, IED.

[67] Art. 18, IED.

- Plant Protection Regulation: in the assessment connected to the approval of active substances, synergists and safeners as well as for the approval of plant protection products,
- Industrial Emission Directive: in some of the BAT conclusions, but not in the Directive itself.

Currently, the WFD considers combination effects explicitly only via the setting of EQS values for groups of chemicals, such as dioxins or brominated flame retardants. However, the ecosystem perspective of the WFD opens up for a more holistic approach: if the aim is to ensure a "good ecological status", then the joint toxicity of all chemicals that an ecosystem is exposed to, need to be taken into consideration.

As presented above the WFD and MSFD creates important links between sector-specific acts and the ecosystem-based approach. BPR provides a legal base for rejecting an application if the intended use of a product endangers the fulfilment of a WFD or a MSFD objective.[68] Similarly, if the use of a plant protection product is endangering the achievements of WFD objectives, targets and EQS, it is possible to withdraw or review already existing approvals for active substances and products under PPR. Assessments and approvals made under these sector-specific acts are thus directly affected if environmental objectives or EQSs under the WFD (and in relation to BPR also MSFD) are derived or updated as a result of a change in the environment. Under REACH an authorisation to use a dangerous substance can be reviewed. This is possible if the use of the substance endangers the fulfilment of WFD art. 4.1, which includes both chemical and ecological status.

BPR and PPR are comparable as both are sector-specific regulations and cover similar types of substances although with different areas of application. Both are more recent than REACH and the specific mixture guidance documents provided indicate an increased awareness within the regulatory bodies of the EU of possible threats from combinations of chemicals.

REACH largely fails to realistically address environmental pollution from mixtures of chemicals as it is based on the current assessment paradigm, performing a substance-by-substance assessment of substances produced and imported at high volumes. In general, it does not consider combination effects, except in relation to MCS and UVCB substances. REACH had, as a draft, an assessment tool for the combined effects of mixtures, which was removed from the final version. However, since 2010 a new combination consideration has been included when compiling a Safety Data Sheet and denoting the combined effects of several components in a mixture. This requirement only considers effects in the human body but indicates that regulators are now willing to address the risks of combination effects in the context of REACH.

---

[68] Annex VI (paras. 67 and 69) of BPR.

The IED lacks combination considerations in the directive itself. Nevertheless, the construction with BAT conclusions provides a branch specific description of the current standard regarding both technological development and environmental concerns. This increases the flexibility to include combination considerations in each BAT conclusion, if appropriate and needed.

International law relating to the environmental status of the Baltic Sea does not explicitly address the issue of chemical mixtures, except in very general terms that require measures to be taken to avoid pollution and an unhealthy environment. However, the BSAPs relation to the MSFD gives not only a stronger enforcement of the objectives in BSAP in relation to hazardous substances, but also promotes the consideration of the combination effects of different types of pressures on the marine environment.

5.6 Conclusions

Even if the impact of each individual chemical emission into the Baltic Sea might be acceptable, their combined effects might not. This is a major shortcoming in international and EU legislation, i.e. that chemicals are assessed and managed based on the notion that each chemical is the only toxic substance present in an otherwise pristine environment. This is a dangerous oversimplification of reality that no national legislation seems to be able to deal with. It is not sufficient to ensure that the actions of an individual company, authority or consumer are on an acceptable level – it is their joint impact that matters when ensuring good environmental status. The failure to set boundaries for the total chemical impact on the Baltic Sea from multiple chemicals, sources and activities erroneously implies an open, infinite ecosystem. This obviously incorrect assumption is the Achilles heel of current chemical regulation relating to the Baltic Sea.

The current chemical status of water and sediments in the Baltic Sea is an illustration of chemical policies over the last 50 years. The few examples of decreases in substances like DDT present a positive response to legislative initiatives, while the above mentioned classification of the most part of the Baltic as moderate, bad and poor reveals insufficient legislation and poor policy outcomes.[69] This cannot be a lack of knowledge about the situation, it is a lack of political will to favour concern about the environment and public health over short-term economic growth. It is also important to recognise that past pollution loads, although the original source is no

---

[69] The production, trade and use of DDT was banned between 1970 and 1975 in most countries bordering the Baltic Sea, notably in Scandinavia and the former West Germany. DDT was no longer used in Poland by the mid-1980s, nor in Estonia, Latvia, Lithuania, Russia and the former East Germany by the mid-1990s. See HELCOM, 2010. Hazardous substances in the Baltic Sea: An integrated thematic assessment of hazardous substances in the Baltic Sea. Baltic Sea Environment Proceedings No. 120B, 32–36.

longer present in water, might be buried in soils or sediment that has not yet disappeared from the ecosystem.

EU chemical legislation, although not sufficient to handle chemical mixtures, has developed to increasingly consider the combination effects of chemicals and uses a more precise terminology. The examined regulations and directives can be viewed as belonging to two different generations with respect to the consideration of the combination effects expressed in their terminology. The older generation is represented by WFD and REACH, which both derive from the late 1990s or early 2000s. They use different terms for expressing combination effects. The newer generation is PPR, BPR and MSFD, which explicitly uses the more precise terms *cumulative and synergistic effects* (in MSFD *synergetic*).

There are cross-regulatory mechanisms in place, linking the application of product-related regulation like REACH, BBR and, PPR to the environmental status of the Baltic Sea. These links could be expanded with provisions requiring compliance with the environmental objectives of the WFD and MSFD for both the authorisation and approval of substances, and the withdrawal of authorisation to use a substance. These links are still far too weak and insufficient as long as there are no objectives or EQS relating to mixture toxicity. Developing toxicity EQS under WFD or MSFD would have an effect on the application of approval and review tools in sector-specific acts. The consideration of mixture effects in all type of risk assessments under REACH, BPR and PPR would also be a way to deal with the risks of mixture effects.

Chapter 6
Salvage of Wrecks in the Baltic Sea – A Finnish Perspective

Jan Aminoff

Abstract The chapter addresses legislation on the salvage of wrecks. It addresses relevant international and national rules and principles regarding protection of underwater cultural heritage and case law concerning salvage of wrecks and historical wrecks, with the aim of identifying possible gaps in the statutory protection of historical wrecks. Certain examples of how these issues have been addressed in the other Baltic Sea states are provided. The chapter also comments on the rights and obligations of the owners of wrecks and their cargos, on the one hand, and those of the salvors, on the other. The main emphasis is on Finnish law, but many observations are equally valid in other Baltic Sea countries.

Keywords Salvage · Wrecks · Historical wrecks · Underwater cultural heritage · Baltic Sea · Finland

6.1 Introduction

The focus of this chapter is on salvage of wrecks. We only have to take a look at the sea charts covering the coasts of the states surrounding the Baltic Sea, including the Gulf of Finland and the Gulf of Bothnia, to understand that the salvage of vessels is an important activity in this region. For example, the sea charts of the Finnish and Swedish coasts disclose that the waters are difficult to navigate due to numerous islands, narrow fairways and rocks both above and below sea level. According to the Finnish National Board of Antiquities and Historical Monuments, some 1500 shipwrecks have been located in Finnish territorial waters of which approximately 660

Former attorney having practised maritime and transport law for some 38 years; LL.Lic, trained on the bench.

J. Aminoff (✉)

are ships which sank more than 100 years ago.[1] Along the Swedish coast at least some 4000 remains of shipwrecks and similar objects have been registered, many of which are historical wrecks. Most of these wrecks are located in Swedish territorial waters.[2] But the existence of wrecks is of course not limited to the Finnish and the Swedish coasts. According to one source there are about 80,000 vessels at the bottom of the Baltic Sea that sank during the past 600 years, which renders the Baltic Sea unique in this respect.[3]

A number of international conventions and national legislation regulate salvage law and the performance of salvage operations in the Baltic Sea. Some of these conventions have been ratified by the Baltic Sea Coastal States (BSCS)[4] and also one by the European Union. The most important convention is the International Convention on Salvage of 1989 (Salvage Convention 1989). This has been ratified by all the BSCS, including the Russian Federation. The last of the BSCS to ratify the convention was Finland, where the convention entered into force on 12 January 2008. The European Union is not a party to this convention. This convention constitutes the backbone of salvage law in the Baltic Sea region.

In addition, the 1976/1996 Convention on Limitation of Liability for Maritime Claims (LLMC) is important since it entitles salvors to limit their liability for services rendered in direct connection with salvage operations.[5] This Convention has been ratified by all the BSCS but will not be further considered in this presentation.

The United Nations Convention on the Law of the Sea (UNCLOS) of 1982 has been ratified by all the BSCS including the European Union and the Russian Federation. The relevance of this convention to salvage law relates to its provisions that deal with the underwater cultural heritage, such as old wrecks lying on the seabed (arts. 149 and 303).

The 2001 UNESCO Convention on the Protection of the Underwater Cultural Heritage (UCH Convention) entered into force internationally on 2 February 2009. As of 20 January 2016, the UCH Convention had been ratified by 54 countries, including Lithuania, but not by any of the other BSCS. The European Union is not a party to this Convention. So far, Finland has not taken any steps towards ratification. Art. 4 contains a provision which, in principle, excludes the application of salvage law and the law of finds to the underwater cultural heritage to which the Convention applies. Underwater cultural heritage in this context means vessels, air-

---

[1] J. Aminoff *Historic Wrecks and Salvage under Finnish Law – Recent Developments* in H. Rak and P. Wetterstein (eds) *Shipwrecks in International and National Law* (Institute of Maritime and Commercial Law – Åbo Akademi University, 2008) 116; Information received by e-mail in May 2016 from the Finnish National Board of Antiquities.

[2] Swedish National Heritage Board, Fornsök.

[3] B. Grisell and L. Randall *Vårt marina arv* (available at https://www.abc.se/-pa/publ/grisells.htm) 1.

[4] Finland, Estonia, Latvia, Lithuania, the Russian Federation, Poland, Germany, Denmark and Sweden.

[5] Convention on Limitation of Liability for Maritime Claims, 1976, and the Protocol of 1996 to amend the Convention on Liability for Maritime Claims of 19 November 1976.

craft, other vehicles or any part thereof, their cargo or other contents, together with their archaeological and natural context, which have been partially or totally lost under water, periodically or continuously, for at least 100 years (art. 1. (a)).

The Nairobi International Convention on the Removal of Wrecks of 2007 (WR Convention) entered into force internationally on 14 April 2015. By 31 October 2016 the convention had been ratified by Denmark, Finland and Germany, but not by any of the other BSCS. The European Union is not a party to this convention. Sweden and Norway are also preparing for ratification. The WR Convention concerns, firstly, the obligation of the contracting States to take measures in relation to the removal of wrecks that pose a hazard in the convention area, and, secondly, the obligation of the registered owner of these wrecks to comply with instructions given by the state affected. This convention does not directly concern the relationship between salvors and the owners of wrecks and their cargo. However, it might, at least in theory, have an effect on the performance of a salvage operation in that a contracting State is entitled to lay down conditions for the performance of salvage operations.

National legislation in each BSCS, such as the Maritime Codes and other similar legislation, will contain provisions on salvage law and other questions relevant to the salvage of wrecks. In Finland the provisions of the Salvage Convention have been incorporated into the national Maritime Code. This is also the case in at least Sweden.

No EU legislation in force directly regulates salvage law. However, a number of EU legislative acts exist relating, for example, to pollution of the environment, maritime safety and the protection of the marine environment. These laws may have a direct or indirect effect on the performance of a salvage operation.[6] This legislation will not be dealt with in this presentation.

In conclusion, it can be said that salvage law in the BSCS is governed by the 1989 Salvage Convention, by national law and by those provisions of the 1982 UNCLOS that concern underwater cultural heritage. Since the European Union is also a party to the UNCLOS, this convention forms part of EU law. Lithuania is so far the only BSCS which has ratified the UCH Convention. The WR Convention has been ratified by Denmark, Finland and Germany but not yet by the other BSCS.

This chapter will a) analyse the impact of relevant international conventions and principles regarding protection of underwater cultural heritage on Finnish national law and case law concerning salvage of wrecks and historical wrecks, and b) aim at identifying possible gaps in the statutory protection of historical wrecks. Indicative comments c) regarding how these issues have been solved in the other BSCS will also be made. The chapter will also d) include comments on the rights and obligations of the owners of wrecks and their cargos, on the one hand, and those of the salvors, on the other. The comments on Finnish law are also presumed relevant in the other BSCS, at least those comments that are based on the Salvage Convention.

---

[6]A. Rosas "Shipwrecks in EU Law" in H. Rak and P. Wetterstein (eds) *Shipwrecks in International and National Law* (Institute of Maritime and Commercial Law – Åbo Akademi University, 2008) 43–50.

But first, some comments will be made on the relevance of the UNCLOS to salvage. Corresponding views will also be expressed on the relevance of the UCH Convention and the WR Convention assuming that these Conventions will one day enter into force in all the BSCS. The Salvage Convention will be dealt with more extensively in the section entitled 'Scope of application of the Salvage Convention and the Nordic perspective' below. Before the presentation of the actual subject matter some words will be said about the old, sometimes controversial, relationship between salvors and those who want to protect underwater cultural heritage from being touched by salvors.

6.2 Salvage Law Versus Protection of Underwater Cultural Heritage

The origin of the law of salvage is of great antiquity and is to be found in ancient legal systems. During times when looting of goods found in or originating from shipwrecks was frequent, the purpose of introducing the right to a salvage reward is said to have been to prevent or discourage people from committing such acts. Since then the policy of encouragement to salvage is held to be one of the fundamental principles behind the law of salvage. The prospect of obtaining a salvage reward has and will continue to inspire salvors to locate ships and wrecks and to invest funds and use equipment of their own in raising and salving these vessels.

The "no cure-no pay" principle is also an essential element in salvage law; in order to be entitled to a salvage reward, the salvage operation must have been successful. The risk of failure is on the salvor.[7]

The importance of activities by private salvors in locating and raising wrecks and old historical wrecks should not be underestimated. Individual States often lack sufficient financial means and capacity to carry out such operations. It has been said that divers are one of the main contributors to new discoveries of underwater cultural objects. However, divers are also held to pose a threat to the preservation of underwater cultural objects for future generations.[8]

It has been argued that the greatest museum of human civilisation lies on the seabed. Scovazzi explains that today the capacity of some States and private entities to use advanced technology to explore the seabed at increasing depths not only allows access to a huge cultural heritage, but also entails the risk of that heritage being looted or used for private gain.[9] Sometimes the activities of salvors have been

---

[7] Aminoff, note 1 at 115.

[8] B. Varenius, *RUTILUS: Strategies for a Sustainable Development of the Underwater Cultural Heritage in the Baltic Sea Region* (Swedish Maritime Museums Report dnr 1267/03–51, 2006) 5.

[9] T. Scovazzi, "Protection of the underwater cultural heritage" in R. Caddell and D.R. Thomas (eds) *Shipping, Law and the Marine Environment in the twenty-first Century* (Lawtext Publishing Limited, 2013) 293.

compared to a "freedom-of-fishing approach based on exploiting underwater cultural heritage for the purpose of private interests and for the finder's personal gain".[10] Thus, tension exists between salvors and those wanting to protect underwater cultural heritage by preventing salvors from touching and engaging in the raising of such objects.

The United Nations General Assembly noted in its Resolution 66/231 of 24 December 2001 "that underwater archaeological, cultural and historical heritage, including shipwrecks and watercraft, holds essential information on the history of humankind and that such heritage is a resource that needs to be protected and preserved". This traditional approach expressed in the UN General Assembly resolution has led to several national and international initiatives, such as UNCLOS and the UCH Conventions being adopted and ratified by many States. UNCLOS does not specify any criteria for when a sunken ship qualifies as an archaeological and historical object. However, the UCH Convention sets the limit at 100 years.

Although Finland is not a party to the UCH Convention, Finnish national law contains provisions to the effect that vessels and wrecks which sank to the bottom of the sea at least 100 years ago and their cargo are protected and cannot be the object of any salvage operations without the authorisation of the Finnish authorities. The same principle of protection applies at least with regard to Sweden, Norway and Denmark. All other BSCS also have provisions in their national law protecting historical wrecks. In this chapter vessels that sank at least 100 years ago are referred to as historical wrecks, but the notion also covers wrecks protected under the respective national laws of the other BSCS regardless of which yardstick applies to this concept. For example, Sweden used to apply the 100 year yardstick but as of 1 January 2014 a historical wreck is a vessel that sunk before 1850.[11] The 100 year yardstick is also applied in some of the other BSCS. In Estonia, no age limit seems to apply, but wrecks that can be classified as a cultural object of special interest are protected, while some of the other countries apply an age limit between 30 and 50 years. Denmark allows the protection of wrecks younger than 100 years – under special circumstances.[12] In this presentation protection of wrecks is assumed to mean that salvage operations are excluded unless permission has been obtained from the local authorities.

---

[10] Ibid., 295.

[11] Swedish Historic Environmental Act (1988:950) ch. 2 arts. 1(a), 4(2) and 6.

[12] Varenius, note 8 at 18–19, Appendix I 1.1.

6.3   Short Review of Some International Conventions
        Relevant to Salvage

6.3.1   The 1982 United Nations Convention on the Law
            of the Sea

Only two provisions of the UNCLOS concern underwater cultural heritage, namely arts. 149 and 303. Art. 149 states that:

> All objects of an archaeological and historical nature found in the Area shall be preserved or disposed of for the benefit of mankind as a whole, particular regard being paid to the preferential rights of the State or country of origin, or the State of cultural origin, or the State of historical and archaeological origin.

This specific provision applies to the Area, that is, the seabed and the ocean floor beyond the 200 nautical mile limit.[13] The provision is vague. Moreover, it does not specify how it should be applied in practice or the contents of the rights referred to. Nor does it define the concept, "[A]ll objects of an archaeological and historical nature". However, it evidently includes old ships/wrecks at the bottom of the sea. It has been stressed that this provision gives little weight to private interests, such as using found salved objects for trade and private gain. Instead, the emphasis is on preservation and disposal of these objects for "the benefit of mankind as a whole".[14]

Art. 303 concerns archaeological and historical objects found at sea. The article reads as follows:

1. States have a duty to protect objects of an archaeological and historical nature found at sea and shall cooperate for this purpose.
2. In order to control traffic in such objects, the coastal State may, in applying article 33,[15] presume that their removal from the seabed in the zone referred to in that article without its approval would result in an infringement within its territory or territorial sea of the laws and regulations referred to in that article.
3. Nothing in this article affects the rights of identifiable owners, the law of salvage or other rules of admiralty, or laws and practices with respect to cultural exchange.
4. This article is without prejudice to other international agreements and rules of international law regarding the protection of objects of an archaeological nature.

Art. 303(1) establishes a general obligation to protect and cooperate. This obligation applies to all archaeological and historical objects, wherever they are found at sea, including finds in internal maritime waters, archipelagic waters and the 12-mile

---

[13] Art. 1 (1) of the UNCLOS defines the Area as follows: "Area means the seabed and ocean floor and subsoil thereof, beyond the limits of national jurisdiction." This means the seabed and ocean floor beyond the 200 nautical mile limit.

[14] Scovazzi, note 9 at 298.

[15] Art. 33 of the UNCLOS concerns the right of a coastal State to exercise control to prevent infringements of its customs, fiscal, immigration etc. laws within its territory or territorial sea and to punish infringements of such laws.

territorial sea of a state. A state that knowingly destroys or allows the destruction of objects belonging to underwater cultural heritage can be held responsible for a breach of the obligation to protect.[16]

Under art. 303 (2) a special regime applies to archaeological and historical objects located within the 24-mile contiguous zone. The interpretation of this provision seems unclear. Moreover, its application in practice may lead to complications, since art. 33 formally only concerns the infringement of local customs, fiscal, immigration or sanitary law.[17]

Scovazzi points out that specific provisions apply to the space between 12 and 24 nautical miles and to the Area but that UNCLOS does not define a regime relating to the underwater cultural heritage found in the space located between the external limit of the archaeological zone and the Area, that is, between 24 and 200 nautical miles. According to Scovazzi, the legal vacuum left by UNCLOS greatly threatens the protection of cultural heritage since it introduces the abstract idea of the freedom of the seas. According to this principle of freedom, any person on board a ship could explore the continental shelf adjacent to any coastal State, bring archaeological and historical objects to the surface, become their owner under domestic legislation, carry objects to certain countries and sell them on the private market.[18]

According to Scovazzi the risk of uncontrolled activities is further aggravated by art. 303 (3). This subjects the general obligation to protect archaeological and historical objects to the exclusion of the rights of identifiable owners and the law of salvage and other rules of admiralty. In case of a conflict between the obligation to protect underwater cultural heritage under art. 303 (1) and the exceptions under art. 303 (3), the latter will prevail.[19]

The UNCLOS does not define the concept of "the law of salvage and other rules of admiralty". Scovazzi explains that in many countries the notion of salvage is only related to attempts to save a ship or cargo from imminent marine perils on behalf of its owners. That is, it was never intended to apply to submerged archaeological sites or sunken ships which, far from being in peril, have definitely been lost. On the contrary, says Scovazzi, in a minority of common law countries, particularly in the United States, the concept of salvage law has been expanded by some court decisions to cover activities which have very little to do with traditional salvage.[20] In this context it is worth mentioning that the Nordic Maritime Codes extend the concept of salvage beyond the formal scope of the Salvage Convention, which implies that the vessel is in danger, to include vessels and wrecks that are not in immediate danger but have foundered.[21]

---

[16] Scovazzi, note 9 at 296–297.

[17] Ibid., 297; H. Ringbom, *"Wrecks in International Law"* in H. Rak and P. Wetterstein (eds) *Shipwrecks in International and National Law* (Institute of Maritime and Commercial Law, Åbo Akademi University, 2008) 25.

[18] Scovazzi, note 9 at 298.

[19] Scovazzi, note 9 at 299.

[20] Ibid.

[21] FMC ch. 16 art. 1 (1).

Finally, according to art. 303(4), art. 303 is without prejudice to other international agreements and rules of international law regarding the protection of objects of an archaeological nature. This provision seems to leave room for conventions, such as the UCH Convention, which explicitly excludes the application of salvage law and the law of finds to underwater cultural heritage. But as long as the UCH Convention does not apply, UNCLOS does not seem to prevent BSCS from applying salvage law to underwater cultural heritage. However, provisions preventing salvage law from applying to underwater cultural heritage may have been incorporated in the national laws of the BSCS. This is a question that will be further considered in this presentation.

6.3.2   The 2001 Convention on the Protection of the Underwater Cultural Heritage

As mentioned before, of the BSCS only Lithuania has ratified the UCH Convention to date. Finland has no immediate plans to ratify, but it appears that UNESCO will arrange a meeting in the near future to discuss greater support for this convention.

It is said that the legal framework of this convention stems from two distinct sections of UNCLOS, including arts. 149 and 303.[22]

The UCH Convention aims at ensuring and strengthening the protection of underwater cultural heritage (art. 2 (1)). The states that are party to the UCH Convention are to preserve underwater cultural heritage for the benefit of humanity in conformity with the provisions of the convention (art. 2 (3)). Recovered underwater cultural heritage should be deposited, conserved and managed in a manner that ensures its long-term preservation (art. 2 (6)). Underwater cultural heritage is not to be commercially exploited (art. 2 (7)).

Art. 1 prescribes that for the purpose of the UCH Convention:

1. "Underwater cultural heritage" means all traces of human existence having a cultural, historical or archaeological character which have been partially or totally under water, periodically or continuously, for at least 100 years such as:

    (i) sites, structures, buildings, artefacts and human remains, together with their archaeological and natural context;
    (ii) vessels, aircraft, other vehicles or any part thereof, their cargo or other contents, together with their archaeological and natural context; and
    (iii) objects of prehistoric character.

In contrast with UNCLOS, the UCH Convention contains a definition of underwater cultural heritage. This includes a useful yardstick to apply when assessing how long an object must have been partially or totally under water to qualify as

---

[22] https://en.wikipedia.org/wiki/UNESCO_Convention_on_the_Protection_of_the_Underwater_Cultural_Heritage

underwater cultural heritage under the convention. The requirement is at least 100 years. The convention also expressly refers to vessels, aircraft, or other vehicles or any part thereof and their cargo or other contents. This reference is lacking in UNCLOS.

In line with the provision in art. 2(7) that underwater cultural heritage is not to be commercially exploited, art. 4 states that any activity relating to underwater cultural heritage to which the convention applies will not be subject to the law of salvage or of finds, unless it

(a) is authorized by the competent authorities, and
(b) is in full conformity with this convention, and
(c) ensures that any recovery of the underwater cultural heritage achieves its maximum protection.

Although the convention does not totally exclude the application of the law of salvage and finds, in practice the convention prevents all the effects, often considered undesirable, of applying salvage law to underwater cultural heritage. Scovazzi concludes that "freedom to fish" for archaeological and historical objects underwater is definitely banned.[23] Indeed, art. 4 also seems to eliminate all the traditional incentives for salvors to engage in salvage operations concerning underwater cultural heritage that has been under water for more than 100 years.

Art. 3 of the UCH Convention states that nothing in the convention will prejudice the rights, jurisdiction and duties of States under international law, including UNCLOS. The UCH Convention is to be interpreted and applied in the context of and in a manner consistent with international law, including UNCLOS. Art. 303 (4) of UNCLOS contains a similar provision, stating that art. 303 does not prejudice other international agreements and rules of international law regarding protection of objects of an archaeological and historical nature. Art. 303 (3) of UNCLOS does not place any restrictions on salvage law while the UCH Convention contains provisions that in practice eliminate the application of salvage law to underwater cultural heritage. In so far as a conflict arises between these two conventions, the UCH Convention will probably prevail on the basis of the *lex specialis* and *lex posterior* principles.[24]

The parties to the UCH Convention have an exclusive right to regulate and authorise activities directed at underwater cultural heritage in their internal waters, archipelagic waters and territorial sea (art. 7(1)). The convention also contains provisions covering the contracting parties' rights and obligations when regulating and authorising activities directed at underwater cultural heritage in the contiguous zone (art. 8), the exclusive economic zone and on the continental shelf (arts. 9, 10) and in the Area (arts. 11, 12).

---

[23] Scovazzi, note 9 at 301.
[24] Ringbom, note 17 at 30–31.

6.3.3   The Nairobi International Convention on the Removal of Wrecks of 2007

The convention has been ratified by Denmark, Finland[25] and Germany. Sweden and Norway are preparing to join. The convention contains provisions regarding the removal of wrecks that pose a hazard or impediment to navigation or may reasonably be expected to result in major harmful consequences to the marine environment, or damage to the coastline or related interests of one or more states. The convention concerns wrecks following upon a maritime casualty as defined in art. 1.3 of the convention, not operational vessels which have been abandoned.[26] The convention sets out rules, inter alia, regarding defining a wreck (art. 1.4), reporting wrecks (art. 5), determination of hazard (art. 6), locating wrecks (art. 7), marking of wrecks (art. 8) and measures to facilitate the removal of wrecks (art. 9). It is for the state concerned to determine whether a wreck poses a hazard in the Convention area and take measures in relation to the removal of such a wreck (arts. 2.1 and 6). The registered owner must remove a wreck which poses a hazard to navigation and the owner is also liable for the costs of locating, marking and removing the wreck (arts. 9.2 and 10.1). The registered owner may contract with any salvor or other person to remove a wreck determined to constitute a hazard. Before removal commences, the state concerned may lay down conditions for removal (art. 9.4). It is of interest to note that, to the extent that measures taken under the convention are considered to be salvage under applicable national law or an international convention, such a law or convention will apply to questions of remuneration or compensation payable to salvors to the exclusion of the rules of the WR Convention (art. 11.2).

The WR Convention is applicable in the exclusive economic zone of a contracting state, which is an area beyond and adjacent to the territorial sea, which shall not extend beyond 200 nautical miles from the baseline from which the breadth of the territorial sea is measured.[27] However, a state party may extend the application of the WR Convention to wrecks located in its territory, including the territorial sea (art. 3.2). At least Finland has availed itself of this possibility, and the intention of the Swedish legislator is to extend the scope of the WR Convention to their territorial sea.[28]

The WR Convention does not seem to have a direct impact on salvage law. However, since the State concerned has a right to impose conditions for the removal of a wreck, the convention may affect the performance of a salvage operation. On the other hand, it should be reassuring for salvors to note that the convention does not affect the traditional method of assessing the salvage reward.

---

[25] The Convention was ratified by Finland on 27 October 2016 and entered into force three months later.

[26] B. Bengtsson *Avlägsnande av. vrak* Ds 2015:16 (Stockholm, 2015) 44–45.

[27] Arts. 55 and 57 of UNCLOS; art. 1.1 of the WR Convention.

[28] Finnish Draft Government Bill concerning the adoption of the WR Convention 1–2, Bengtsson, note 26 at 4.

6.4 Definition of a Shipwreck

Before turning to the principles of salvage law applicable to wrecks it is justified to commence by defining the concept of a shipwreck. Neither the Salvage Convention nor the FMC contains a definition of a shipwreck. However, the jurisprudence contains several definitions of this concept. For example, Hugo Tiberg concludes that under Swedish law a wreck is "a vessel that is or has become so damaged or leaking that there is none or little prospect of its salvation, as well as floating parts of a vessel or its rigging". In Anglo-Saxon law, says Tiberg, a wreck is said to be "a unit not capable of navigation", and not even a sunken vessel need necessarily be counted as a wreck if it can be salvaged to be navigated again.[29] Tiberg refers to a Particular Average Statement published in Nordiske Domme (ND) in 1990 in which the Swedish Average Adjuster discussed whether the Swedish Ro/Ro vessel *Vinca Gorton* was a wreck when she had sunk in 25 metres' depth off the Dutch coast and had probably been broken asunder. In general parlance she would certainly be regarded as a wreck, said the adjuster, although he would not classify her as such so long as the possibility existed of salving her.[30] It should be emphasized that the Average Adjuster considered the status of the sunken ship from an insurance law perspective.

Peter Wetterstein concludes that the notion of a shipwreck generally includes vessels that are difficult to salvage, in particular, if they have been in the water for a longer time. In this context, Wetterstein also refers to the *Vinca Gorton* case.[31] He also concludes that a foundered vessel in the meaning referred to in the FMC, as a starting point, also comprises old shipwrecks.[32]

The 2007 Convention on the Removal of Wrecks, which has entered into force in Finland, Germany and Denmark and will before long also enter into force in Sweden, contains a definition of a wreck in art. 1(4): "Wreck", following upon a maritime casualty, means:

(a) a sunken or stranded ship; or
(b) any part of a sunken or stranded ship, including any object that is or has been on board such a ship; or
(c) any object that is lost at sea from a ship and that is stranded, sunken or adrift at sea; or
(d) a ship that is about, or may reasonably be expected, to sink or to strand, where effective measures to assist the ship or any property in danger are not already being taken.

---

[29] H. Tiberg *Wrecks and Wreckage in Swedish Waters* (published on the website of the institute of Maritime Law and other Transport Law in 2004) 3.

[30] *Ibid.*; *Nordisk Dommesamling i Sjöfartsanliggender 1990 (ND)*, 8.

[31] P. Wetterstein *Redarens miljöskadeansvar* (Åbo Akademis förlag - Åbo Akademi University Press, 2004) 323–324.

[32] P. Wetterstein "Vrak och gamla skatter" (2000) 2000:5 *Juridiska föreningens tidskrift (JFT)* 451–484, at 456.

The approach to the definition of a wreck under the WR Convention is different from the traditional definitions expressed in civil law jurisprudence. This is evidently due to the fact that the object of the convention is to eliminate hazards to navigation or the environment posed by wrecks. In the first place, the Convention concerns the rights and obligations of the contracting States and does not address the relationship between a salvor and the owner of a wreck and its cargo. The definition of a wreck contained in the Convention will most likely influence the general perception of a shipwreck if and when the Convention becomes part of national law.

In addition to sunken or stranded vessels the concept of a wreck under the WR Convention also includes objects that are or have been on board a sunken or a stranded vessel, as well as objects originating from a vessel but lost at sea which have stranded, sunk or are adrift at sea. The notion of a wreck is further extended to ships that are about, or may be reasonably expected, to sink or strand. The definition of a wreck contained in the WR Convention is thus fairly, if not very, wide.

6.5   Scope of Application of the Salvage Convention and the Nordic Perspective

The preamble to the 1989 Salvage Convention concludes, inter alia, that the States Parties to the convention are: a) "[C]onscious of the major contribution which efficient and timely salvage operations can make to the safety of vessels and other property in danger and to the protection of the environment," and b) "[C]onvinced of the need to ensure that adequate incentives are available to persons who undertake salvage operations of vessels and other property in danger".

These concerns are partly reflected in art. 1 (a) stating that for the purpose of the convention:

(a) Salvage operation means any act or activity undertaken to assist a vessel or other property in danger in navigable waters or in any other waters whatsoever.

Consequently, for the provisions of the 1989 Salvage Convention to apply, a vessel and other property assisted by salvors should be in danger. Under the Convention a vessel means any ship or craft, or any structure capable of navigation (art. 1 (b)) and property means any property not permanently and intentionally attached to the shoreline and also includes freight at risk (art. 1 (c)). However, the Convention does not apply to fixed or floating platforms or to mobile offshore drilling units when on location and engaged in the exploration, exploitation or production of seabed mineral resources (art. 4).

As mentioned before, the provisions of the 1989 Salvage Convention have been incorporated into the Finnish Maritime Code (FMC). Additionally, however, the scope of the Maritime Code has been extended to include salvage of foundered vessels as well. The relevant provision is found in ch. 16 s. 1 (1) of the FMC. The term salvage means "any measure undertaken to assist a vessel or other property which has foundered or is in danger in navigable waters". A vessel can founder without being in danger. A vessel has foundered when she has sustained damage at sea and is no longer afloat. An example would be when she has sunk or is stuck on a rock which prevents her from continuing her voyage. In such a situation a vessel does not have to be in danger for the FMC salvage provisions to apply. A vessel that sank to the bottom of the sea a long time ago but is still in good shape can hardly be held to be in danger although she has foundered. On the other hand, a vessel can be both foundered and in danger at the same time. To illustrate, she has run aground and the risk is that she will suffer further damage due to an approaching storm.[33]

The other Nordic countries, Sweden, Norway and Denmark, have also extended the scope of their legislation on salvage law to include foundered vessels.

The definition of property in the Salvage Convention includes freight at risk. No such express reference appears in the FMC but freight at risk is undoubtedly covered by the Code. The explanation is that the legislator held that this already follows from prevailing salvage law practice and considered it unnecessary to include a specific reference to that effect in the Code.[34]

In the FMC the notion of a vessel is defined as a conveyance or appliance capable of navigation (art. 1(2)). Although this wording could be held to differ from the corresponding wording of the 1989 Salvage Convention, the intention was not to change the definition adopted in the Convention. Appliances capable of navigation mean items such as drilling rigs and other mobile platforms. As under the Salvage Convention, the Code does not apply to fixed or floating platforms or to mobile offshore drilling units that are on location engaged in the exploration, exploitation or production of seabed mineral resources (art. 4(4)). But the Code does apply, for example, when those platforms or units founder *en route* to their location.[35]

"Other property" means property not permanently and intentionally attached to the shoreline (arts. 1(1) and (3)). This includes property carried on board or belonging to the vessel and, in addition to property belonging to the crew or passengers, other property not expressly excluded by the definitions of the Code/Convention.[36]

Like the 1989 Salvage Convention, the FMC implies that a salvage operation has been provided in navigable waters. This means both the open sea and inland waters such as lakes, rivers and creeks.[37] Salvage services do not have to be performed from a vessel and can, for example, be provided from a helicopter or by using a shore crane.[38]

The FMC salvage provisions apply to wrecks no longer in danger. This standpoint is supported in jurisprudence[39] and also by case-law. In *Vrouw Maria* the

---

[33] Finnish Government Bill 97/2006 (Adoption of the Salvage Convention and its incorporation in the Maritime Code), at 22.

[34] Ibid., 23.

[35] Finnish Government Bill, note 33, at 24.

[36] Ibid., 22.

[37] Art. 1(a) of the Salvage Convention refers to "… [I]n navigable waters or in any other waters whatsoever".

[38] Finnish Government Bill, note 33, at 22.

[39] Wetterstein, note 32 at 455–456.

Turku Court of Appeal concluded that the FMC salvage provisions applied to the
Vrouw Maria, a vessel which had sunk off the south-western coast of Finland in
1771[40] and was not held to be in danger.

The 1989 Salvage Convention does not mention wrecks, thereby avoiding the
question whether wrecks should be covered by the Convention. This question
evidently caused controversy during the negotiations leading to the Convention.
The Convention is said to be generally inclusive and does not exclude wrecks.
Consequently, if as a matter of construction, wrecks fall within the terms of the
Convention, then it will apply to them. The Convention applies generally to any
property not permanently and intentionally attached to the shoreline. Additionally,
property which has sunk and become attached to the shoreline should be covered
since, even if it has become permanently attached, it is unlikely to have become
attached intentionally. However, property must also be in danger for the Convention
to apply.[41]

Those BSCS which have not, in contrast to the Nordic countries, extended the
application of the Salvage Convention to foundered ships may not be able to apply
salvage law to shipwrecks that are no longer in danger, unless other national law
exists to that effect.

The Salvage Convention is applicable whenever judicial or arbitral proceedings
relating to matters dealt with in the Convention are brought in a State Party (art. 2).
A corresponding provision is contained in the FMC with regard to civil law disputes
brought before a court or arbitral tribunal in Finland (ch. 16.2). This means that the
Convention and the FMC apply to the salvage of vessels and wrecks in Finnish
internal waters and the territorial sea where Finland exercises territorial sovereignty
(UNCLOS Part II s. I art. 2).[42]

6.6   Relevance of Ownership to Wrecks and Their Cargo

The relevance to salvage law of title to a wreck and its cargo will be discussed.
Sometimes it is unclear who owns an old shipwreck which may have been abandoned. The salvors may then try to claim title to both the wreck and its cargo. The
prerequisites for such an acquisition will also be discussed. Otherwise the traditional legal requirements for ownership of vessels or their cargo will not be dealt
with. The considerations will only concern Finnish law unless an express reference
to the laws of another State is made. On the other hand, comments based purely on
the 1989 Salvage Convention should in principle be applicable in all other BSCS
which are bound by the convention. Under Finnish law the following principles
apply to vessels and wrecks.

---

[40] Turku Court of Appeal, Judgment of 23 March 2005 (S 04/2163), 4.

[41] F. D. Rose *Kennedy & Rose Law of Salvage* (Sweet & Maxwell, 2002) 106.

[42] Y. Tanaka *The International Law of the Sea* (Cambridge University Press, 2015) 5–7.

6.6.1   Conclusion of a Salvage Contract

The owner of a vessel or a wreck concludes a salvage contract with the salvors. However, also the master of the vessel has authority to conclude contracts for salvage operations on behalf of the owner of the vessel. The master or the owner of the vessel also has the authority to conclude such contracts on behalf of the owner of the property on board the vessel (FMC ch. 16, art. 3 (2); Salvage Convention art 6 (2)). However, the authority of the master may become a dead letter with regard to wrecks that sank some time ago since the employment of the master would then normally have ceased. Under certain conditions a salvor is entitled to proceed with a salvage operation although no formal salvage contract exists (the right of the first salvor). That issue will be considered later in this presentation.

The owner of a vessel is also entitled to prohibit the performance of salvage, e.g. if the salvor lacks sufficient skills and appropriate equipment for the operation. This right of refusal is also linked to the right of the salvor to perform salvage services even if no formal salvage contract exists. Services rendered notwithstanding an express and reasonable prohibition by the owner or master of the vessel will not give rise to payment of a salvage reward (FMC ch. 16, art. 10 (2); Salvage Convention art. 19). Such a prohibition must not only be express but also reasonable. It should not conflict with the owners and/or the master's obligation to care for the safety of the vessel, its cargo and its crew. The relevant question is whether the prospects of the vessel being able to manage without assistance are good enough to support such a prohibition. Recently, the protection of the environment has also become increasingly important when considering the reasonableness of a prohibition. A conflict between private and public interests may then easily arise.[43] The state of danger for the vessel and the risk of environmental damage will then be decisive when evaluating the reasonableness of a prohibition.[44] Old wrecks at the bottom of the sea are very seldom in danger, but some wrecks may still have a cargo of oil or bunkers on board. This may entail the risk of environmental damage and could have an impact on the justification for a salvage operation.

A judgment of the Swedish Supreme Court on 13 January 1983 can illustrate the significance of environmental aspects for the performance of salvage operations. The tanker *Tsesis* ran aground on 26 October 1977 in the Swedish archipelago, hitting a rock which was not marked on the official sea chart. The casualty led to some 600 tons of oil leaking from the vessel and causing considerable oil pollution. The Swedish Maritime Administration commenced a salvage operation to minimise the pollution damage although the master of the vessel had refused to conclude a salvage contract. The Supreme Court ruled, inter alia, that even if the master of the *Tsesis* had announced an express prohibition against a salvage operation, it was

---

[43] Wetterstein, note 30 at 456–457.
[44] T. Falkanger, H. J. Bull, *Sjörett* (7. utgave, Sjörettsfondet Akademisk, Oslo, 2010) 451.

evident that the prohibition was not justified, considering the imminent danger of additional pollution damage occurring after the grounding.[45]

6.6.2 Abandonment

Only an owner still in possession of a ship is entitled to decide who will have the right to salve the ship. This authority remains even if the crew leaves the ship. With regard to old wrecks at the bottom of the sea, the continuing existence of ownership is of importance.[46] An owner can, however, under certain conditions, lose ownership of the vessel. The owner may, without assigning ownership or possession to anyone, abandon the ship to be appropriated by others. Such abandonment does not, under Finnish law, require an express manifestation and can, as is often the case, be implicit.[47]

If the owner expressly resigns ownership, then abandonment is easy to verify. But since no express manifestation is required, abandonment may also be established on the basis of objective circumstances. The possibilities in practice to salvage a wreck can be such a circumstance. If a vessel has been lying at the bottom of the sea in shallow waters for many years without any attempts having been made to lift her, it may fairly soon be possible to conclude that the owner has abandoned the vessel. However, regard has also to be taken as to what possibilities the owner has had in practice to proceed with a salvage operation.[48]

With regard to old sunken wrecks, the time factor is of importance. It has been held that only very long passivity on the part of the owner is of relevance. The owner of a ship and its cargo should be given ample time for salvage. It is also possible that the location of a wreck has been unknown for a long time. It is argued that even in such situations, title should prevail for a reasonable time. The original ownership should cease when, considering the lapse of time, the passivity of the original owner and other relevant circumstances, it is reasonable to conclude that ownership has expired. In turn, this will allow other parties to claim ownership of the wreck.[49] This judicial "formula" is wide enough to allow for a considerable amount of disagreement.

The Finnish courts have very seldom dealt with questions relating to the loss of title to old shipwrecks. One such case concerns the Swedish vessel *Jönköping* which was torpedoed in 1916 by a German submarine and sank off the Finnish west coast. The cargo on board consisted partly of a large quantity of bottles of champagne. The wreck was found by a group of divers on 27 May 1997 lying on the seabed 63

---

[45] ND 1983, 1.

[46] Wetterstein, note 32 at 456.

[47] Wetterstein, note 32 at 461; E. Havansi, *Omistusoikeuden hylännästä* (Juhlajulkaisu Simo Zitting 1915–14/2–1985) 68.

[48] Wetterstein, ibid., 462–463.

[49] Wetterstein, ibid., 463–464; Sjur Braekhus, *Tidsskrift for Rettsvitenskap* 1975, 518.

metres below sea level some 11 nautical miles west of the town of Rauma. When the find had gained publicity another group of divers arrived on the scene claiming rights to the cargo and the divers that found the wreck had to resort to protective measures. The first group of divers argued before the Vakka-Suomi District Court that the Swedish owners of the vessel were unknown/had abandoned the wreck and no one had been actively searching for the vessel. Moreover, they claimed that the salvage technique which would enable the raising of the vessel had been known even before the Second World War. On that basis, they argued that the vessel had been abandoned since no one had attempted to salvage her. On 3 July 1998, the District Court concluded that the owners had abandoned the vessel and its cargo and that the divers had taken possession of it as owners by taking steps to prepare for raising the wreck.[50] The reasoning of the court is not very detailed, but clearly the time that had passed since the casualty, the fact that the owners of the vessel and the cargo were unknown and that no one had previously searched for the vessel were decisive factors. The argument that the salvage technique available even before the Second World War would long ago have enabled the lifting of the vessel may also have influenced the court. However, one has to bear in mind that these conclusions were made by a court of first instance in a matter concerning protective measures and not in a dispute regarding material ownership of the wreck. The legal importance of this decision is probably limited.

Another case is the previously mentioned *Vrouw Maria*, litigated in the Turku District Court and the Turku Court of Appeal in 1999–2005. The *Vrouw Maria* ran aground in October 1771 in the south-western archipelago off the Finnish coast, in an area which today forms part of Finnish territorial waters. She sank a few days later on 9 October 1771 with most of her cargo still on board. The vessel was *en route* from Amsterdam to St. Petersburg and is believed to have been carrying a cargo of valuables, paintings and other art objects belonging to the Russian Empress, Catherine the Great. The whereabouts of the wreck were unknown until June 1999, when a group of divers found it on the sea bed 41 meters below sea level. The dispute concerned, *inter alia,* the right to salvage the vessel and her cargo and the ownership of the vessel. Without going into further detail at this stage, suffice it to say that the courts found that the wreck was abandoned with the consequence that others could in principle claim ownership. Bearing in mind that the wreck had been on the seabed for more than 200 years, the conclusion of the court with regard to abandonment is hardly surprising.[51]

---

[50] Decision of the Vakka-Suomi District Court of 3 July 1998 (98/750).

[51] Judgment of the Turku District Court no 04/4312 of 16 June 2004; Judgment of the Turku Court of Appeal no 787 of 23 March 2005.

6.6.3   Ownership by Appropriation

If a wreck or other property has been abandoned, the finder has a possibility to acquire title to that property on the basis of the legal principles governing appropriation. These principles require that:

(a) the property being the object of appropriation must be abandoned by the owner, and
(b) the finder must have acquired control of the property with the intention to become its owner.[52]

In *Jönköping* the divers had lifted a number of bottles of champagne from the holds and argued that they had taken possession of the wreck. They had also built a frame around the wreck to enable appropriate raising. Further, six lifting bags had been attached to the wreck to facilitate lifting. The divers argued that since the wreck had been abandoned by the original owner it now belonged to them. As mentioned above, the District Court ruled that the divers had begun preparations for raising the wreck and otherwise taken possession of it as owners. Clearly the Court must have accepted that the steps taken by the divers met the requirements for obtaining ownership by appropriation.[53]

In *Vrouw Maria* the divers had marked the position of the wreck by placing a buoy nearby. At the request of the Maritime Museum six items had been lifted. The divers had announced their desire to salvage the vessel but the authorities referred to the vessel as being protected by the Antiquities Act and prohibited the divers from further action. In the dispute which followed, the divers claimed that they had acquired ownership of the abandoned vessel subject to the principles of appropriation. The Turku Court of Appeal stated that appropriation may lead to ownership of an abandoned object, such as the wreck that was discovered. A prerequisite of appropriation is that the party claiming appropriation has gained possession of the object in question. Possession of an object is a factual relationship of power, said the Court, and concluded that the provisions of the Antiquities Act concerning protection prevented the divers from obtaining such factual power and that the divers had not gained ownership of the wreck.[54] The conclusions of the Court with regard to possession by the divers are as such formally logical and the situation differs from that of the divers in *Jönköping*. The dispute demonstrates the tension between private interests and those wanting to preserve cultural heritage.

---

[52] Aminoff, note 1 at 126; J. Kaisto and J. Tepora, *Esineoikeus eurooppalaistuvassa Suomessa* (Lakimiesliiton kustannus 2012) 89, 290.

[53] Decision of the Vakka-Suomi District Court of 3 July 1998 (98/750) 2; Application of the divers attached to the Decision 1–2.

[54] Judgment of the Turku Court of Appeal of 23 March 2005 (S 04/2163), 4–5; Under the Antiquities Act vessels having sunk at least 100 years ago are protected and cannot be touched or disposed of without the consent of the authorities.

Due to an amendment of the Antiquities Act in 2002 (art. 20(2)) it is no longer under Finnish law possible to acquire ownership of abandoned historical wrecks by appropriation. This question will be considered later in this paper.

6.6.4   Lost Property Act

The Finnish Lost Property Act[55] governs a system whereby the finder of lost property may under certain conditions become its owner. Lost property should be reported to its owner, if known, and delivered to the authorities. The owner is entitled to regain their property on payment of a reward to the finder. Unless the owner collects the property within three months, title to the property will pass to the finder, with a few exceptions. However, the Act does not apply to property which has evidently been abandoned. Nor does it apply to property covered by the Antiquities Act or to vessels in danger or having foundered or to property carried on board for which a salvage reward is paid subject to the provisions of the FMC. Under the provisions of the Lost Property Act it should not be possible to require title to a shipwreck. If the wreck is old, it will probably have been abandoned and the lifting of the wreck by salvors will require a salvage reward to be paid. If not abandoned, the salvage of the wreck will likewise imply that a reward is paid.

6.7   General Rights of Salvors

This section will focus on discussing some general rights of salvors. These principles apply to salvage of vessels and wrecks. The right of salvage and the right to obtain a salvage reward are of primary importance to salvors. A salvor may also under certain conditions acquire the right to salvage although no formal salvage contract has been concluded.

6.7.1   Right of Salvage and the Salvage Reward

Once the salvage contract is concluded the salvor has the right of salvage and the owner will be liable to pay a salvage reward to the salvor, but only on condition that the salvage operation has had a successful result. In principle, no payment is due if the salvage operation has not been successful. The reward, exclusive of any interest and recoverable and payable legal costs, should not exceed the salvage value of the vessel and other property (FMC ch. 16, s. 5; Salvage Convention art. 12(1), (2), art. 13(3)). The FMC and the Salvage Convention contain detailed provisions on how to

---

[55] Lost Property Act of 26 August 1988/778.

calculate the reward. The reward should be fixed with a view to encouraging salvage operations, taking into account, inter alia, the following criteria (FMC, ch. 16, s. 6; Salvage Convention art. 13(1)):

(a) the salved value of the vessel and other property;
(b) the measure of success obtained by the salvor;
(c) the nature and degree of the danger;
(d) the skill and efforts of the salvors in salving the vessel, other property and life.

The provisions noted above reflect two of the core principles of salvage law: the principle of encouragement and the no cure-no pay principle. The prospect of receiving a reward which reflects, inter alia, the value of the salved property, the degree of danger of the operation, the skill demonstrated and success achieved is intended to encourage a salvor to invest funds and efforts in a salvage operation. As previously mentioned, the purpose of originally introducing the right to a salvage reward, which could in a given situation be substantial, was to discourage and prevent people from looting property from ships. This principle dates back to ancient legal systems. The no cure-no pay principle allocates the risk of success of the operation to the salvor. This can also be held to encourage a salvor to put their best efforts into the operation to earn their reward.[56] Unless the salvage operation is successful the salvor will receive no reward regardless of how much money they have invested in the operation.

A salvor who has carried out salvage operations in respect of a vessel which by itself or its cargo threatened damage to the environment is entitled to special compensation from the owner and the operator for costs incurred. Special compensation is only payable in so far as this compensation exceeds the salvage reward calculated subject to the criteria established in ch. 16 art. 6 of the FMC / art. 13 of the Salvage Convention. Under certain conditions, special compensation can be increased up to a maximum of 30% of the expenses incurred by the salvor (FMC ch. 16 art. 9(2); art. 14(2) of the Salvage Convention). A salvor who has been negligent and thereby failed to prevent or minimise damage to the environment may be deprived of the whole or part of any special compensation due (FMC ch. 16 art. 9(4); art. 14(5) of the Salvage Convention).

A salvor may be deprived of the whole or part of the payment due to the extent that salvage operations have become necessary or more difficult because of fault or neglect on their part, or if the salvor has been guilty of fraud or other dishonest conduct (FMC ch. 16 art. 10(3); art. 18 of the Salvage Convention).

As security for payment of the salvage reward, the salvor has a maritime lien on the salved vessel and salved cargo (FMC ch. 3 s. 2(5) and 9(1)).[57]

---

[56] Aminoff, note 1 at 115; Wetterstein, note 32 at 454–455.
[57] International Convention on Maritime Liens and Mortgages, art. 4(c).

6.7.2   Right of the First Salvor

It has been held that a salvage reward can be claimed even if no explicit salvage contract exists. For example, if the owner has for one reason or another not wanted to appoint a salvor, or has been prevented from doing so, the first salvor to commence salvage operations will have a right of priority to continue operations. It is, however, not sufficient for a salvor to have e.g. come upon a wreck. The salvor must have commenced salvage operations in order to obtain the right of priority to continue operations. This right of priority entitles the salvor to rebuff other salvors.[58] But the owner may also use their right to prohibit the first salvor from continuing the salvage operation, provided that the prohibition is express and reasonable, e.g. if it is clear that the salvor lacks sufficient skills and equipment to properly perform the operation.

What is required of the first salvor in order to secure the right of priority with regard to a wreck? It has been held that it is not enough to have found the wreck, to have marked its position with buoys and to have declared an intention to salve the wreck and its cargo. On the other hand, it is said, the salvor cannot be required to have sent divers down to the wreck or to have placed hawsers around the wreck. The decisive action will presumably be that the first salvor stations their ship(s) furnished with the necessary salvage equipment on board above the location where the wreck is located and that they are ready to proceed with the operation.[59]

One of the issues at stake in *Vrouw Maria* was the right of the first salvor. The divers that found the wreck in June 1999 had marked the location with one or several buoys and at the request of the Maritime Museum lifted a number of objects from the wreck. They had also indicated their willingness to proceed with the salvage operation and provided the court with a written salvage plan containing details on how the operation was to be performed and what equipment was to be used. No other parties competing with the salvors had announced their interest. The Maritime Museum had prohibited the salvors from proceeding with the operation. The Court of Appeal stated in its judgment of 23 March 2005 that the right to prohibit a salvage operation also concerns the so-called first salvor and concluded that the question at issue was whether or not the State had the right to prohibit the divers from proceeding with the salvage of the wreck and her cargo. The Court ruled that the first salvor can only proceed with salvage against the wish of the owner when the property is in concrete danger, which necessarily requires the commencement of salvage operations. Such a situation was, according to the Court, not present. Taking into account that the wreck and its cargo had been at the bottom of the sea for more than 230 years, that the cargo was in good condition and could be presumed to so remain for decades

---

[58] S. Braekhus *"Bergning"* (Universitetsforlaget, Oslo 1971) 54; Wetterstein, note 32 at 458.
[59] Braekhus, note 58 at 55–56; Wetterstein, *"Fru Maria och bärgningsrätten-Konflikt mellan privata och allmänna intressen"* in Wetterstein *Sjörätt med pålägg – Maritime Law with Extras* (Institute of Maritime Law – Åbo Akademi University 2015) 226–227.

and that the wreck constituted no danger to navigation, the Court ruled that the State was entitled to prohibit the salvage of the wreck and its cargo.

Unfortunately, the judgment of the Court of Appeal does not contain an analysis of the prerequisites of the right of the first salvor, which would have been desirable. But since the Court of Appeal elaborates on this right and rejects the divers' request by referring to the right of the owner to prohibit salvage, it can be presumed that the Court, in principle, accepted the divers' requested priority right. Perhaps unsurprisingly, the judgment has been criticised.[60]

6.8 Salvage of Historical Wrecks

The 1989 Salvage Convention entitles any State adhering to the Convention to reserve the right not to apply the provisions of the Convention when the property involved is maritime cultural property of prehistoric, archaeological or historical interest and is situated on the seabed (art. 30 (1) (d)). When ratifying the convention, Finland made a reservation not to apply the provisions of salvage law to historical wrecks and property. The same reservation has been made by the other BSCS except for Latvia, Lithuania and Denmark. Norway has incorporated this reservation into its Maritime Code, whereas in Sweden corresponding legislation is included in the Historic Environmental Act (1988:950). Denmark did not make a reservation when adhering to the Convention, but has included similar legislation in national law.[61] Lithuania is a party to the UCH Convention banning the application of salvage law to historical wrecks. The Maritime Code of Latvia (art. 254(4)) contains a provision that excludes the application of salvage law to ships and objects which have a cultural historic value.[62]

Since 2008 the FMC ch. 16 art. (4) includes the following reservation with regard to applying salvage law to historical wrecks: "nor are the provisions of this chapter [salvage provisions] applicable to vessels and property referred to in the Antiquities Act (295/1963), unless otherwise agreed." This provision is said to clarify the relationship between salvage law and legislation on cultural heritage. The Government Bill also points out that the Salvage Convention only concerns vessels in danger and that it is highly unlikely that the provisions on salvage would become applicable to historical wrecks at the bottom of the sea. However, Finland decided to make the reservation mentioned in art. 30 (1) (d) of the Convention, as indeed have several States of the European Union.[63]

Art. 20 of the Antiquities Act states:

---

[60] Wetterstein, note 59 at 226–230; Aminoff, note 1 at 124–125.

[61] Finnish Government Bill, note 33, at 24; the Finnish legislation entered into force on 12 January 2008.

[62] Information received on 30 August 2016 from Professor Alla Pozdnakova, Scandinavian Institute of Maritime Law, Oslo.

[63] Finnish Government Bill, note 33, at 24; Aminoff, note 1 at 122.

"Any wreck of a ship or other craft discovered in the sea or inland waters, which can be presumed to have sunk at least one hundred years ago, or part of such a wreck, is protected. The provisions regarding immovable antiquities shall apply where relevant to such a wreck or part thereof.

    If, on the basis of external conditions, it is obvious that the owner has abandoned a wreck or part thereof referred to in subsection 1, it shall belong to the State.

    Items found in a wreck referred to in subsection 1, or which apparently originate from such a wreck, belong to the State without redemption, and the provisions concerning movable antiquities shall otherwise apply where relevant."

In other words, vessels/wrecks that sank at least 100 years ago and items found in such wrecks or originating from such wrecks are protected and cannot be the object of the provisions on salvage law, unless otherwise agreed. Since historical wrecks and their cargoes are protected they are subject to supervision by the National Board of Antiquities and any agreement regarding salvage would require the cooperation of the Board. Acceptance by the Board of a traditional salvage contract subject to ordinary salvage law is highly unlikely ever to be obtained.

A point of interest is that under art. 20 subsection 2 of the Antiquities Act, a wreck that sank at least 100 years ago will become the property of the State only if abandoned by its owner, whereas items found in the wreck or which originate from the wreck will automatically become the property of the State. This seems to mean that if the original owner of a wreck or their successor is identified, they may be in a position to claim ownership to the wreck although the vessel in question sank more than 100 years ago. The cargo would, on the other hand, belong to the state without redemption even if the owner is known. Art. 20 subsection 2 of the Antiquities Act also prevents salvors from obtaining title by appropriation to abandoned historical wrecks.

As a curiosity it can be mentioned that subsection 2 of art. 20 of the Antiquities Act came into force on 1 December 2002 with retrospective effect, while the dispute between the divers and the State regarding the *Vrouw Maria* was pending in the Turku District Court. The divers claimed ownership of the wreck on the basis of appropriation and the State amended the Antiquities Act during the proceedings to improve its own position; a highly effective but very unusual and questionable method available only to very few. Both the District Court and the Court of Appeal ruled that the ownership of the wreck belonged to the State on the basis of art. 20 subsection 2 of the Antiquities Act.[64]

Those BSCS having made the reservation when adhering to the 1989 Salvage Convention will have taken steps subject to their respective legal systems to exclude the application of the provisions of the Salvage Convention to historical wrecks and their cargos. Reservations were not made by Denmark, Latvia and Lithuania. Denmark and Latvia have included corresponding legislation in their national law and Lithuania has ratified the UCH Convention. The previously mentioned RUTILUS Report prepared by the Swedish Maritime Museums contains a table

---

[64]Aminoff, note 1 at 126–128.

including information on the respective national legislations in the BSCS that provide protection of wrecks.[65]

6.9 Concluding Remarks

6.9.1 Relevance of International Conventions

Salvage in the BSCS is governed by the 1989 Salvage Convention and national law enacting the Convention. This Convention does not, as such, place any restrictions on applying salvage law to historical or non-historical wrecks.

In addition, a number of international Conventions are of relevance. The 1982 UNCLOS, which, inter alia, aims at protecting objects of an archaeological and historical nature found at sea, such as old wrecks, excludes the rights of identifiable owners, the law of salvage or other rules of admiralty from this protection. In the case of a conflict between the obligation under the Convention to protect historical wrecks and the law of salvage, the latter will prevail. UNCLOS applies in all BSCS.

The 2001 UCH Convention again excludes the law of salvage or of finds from being applicable to underwater cultural heritage having been partially or totally under water, periodically or continuously for at least 100 years, such as vessels (historical wrecks), aircraft or any part thereof, their cargo or other contents, together with their archaeological and natural context. The competent national authorities may, however, authorise the salvage of such cultural heritage. To date only Lithuania has ratified the UCH Convention. The other BSCS are formally not bound by it.

The 2007 WR Convention contains provisions regarding the removal of wrecks that pose a danger or impediment to navigation or that may reasonably be expected to result in major harmful consequences to the marine environment, or damage to the coastline or related interests. The Convention concerns wrecks following upon a marine casualty. The convention does not have a direct impact on salvage law, but since a State involved may impose conditions for the removal of a wreck, the convention may affect the way a salvage operation is performed. In theory, the Convention could also concern historical wrecks. Denmark, Finland and Germany have ratified it and Sweden and Norway are preparing for ratification. Only Denmark, Finland and Germany are bound by the convention.

In conclusion, it can be said that, on an international level, direct restrictions on the application of salvage law to wrecks in the BSCS are only posed by the UCH Convention in those countries that have adhered to the Convention – at the time of writing only Lithuania. This Convention only concerns historical wrecks. With regard to wrecks younger than 100 years no direct international restrictions are in force. The WR Convention does not have a direct impact on salvage law but may place restrictions on the practical performance of a salvage operation.

---

[65]Varenius, note 8, Appendix I 1.1 Table 1. The information contained in Table 1 may not be up to date with regard to all mentioned countries (e.g. Sweden) but should be indicative.

6.9.2 Salvage of Historical Wrecks

Under the Salvage Convention it is possible to make a reservation not to apply the Convention when the property involved is maritime cultural property of prehistoric, archaeological or historic interest situated on the seabed, such as wrecks. This Convention does not define the concept of such wrecks, which is then left to the discretion of the national legislator. Reservations have been made by all the BSCS except for Latvia, Lithuania and Denmark. On the other hand, Lithuania has ratified the UCH Convention and Denmark has incorporated provisions in its national law excluding the collection of a salvage reward concerning maritime cultural heritage.[66] The Latvian Maritime Code contains a provision excluding the application of salvage law to ships and objects which have a cultural historic value.

It may be of interest to note that the Salvage Convention concerns vessels in danger. How is a reservation not to apply the convention to maritime cultural property (historical wrecks) to be understood? Does such a reservation concern only maritime cultural property in danger or any maritime cultural property? Historical wrecks having sunk a long time ago are presumably very seldom in danger. A literal interpretation of the Salvage Convention could mean that historical wrecks not in danger fall outside the scope of the Salvage Convention and would thus be subject to national salvage law, if any, even if a reservation mentioned in art. 30(d) of the Convention has been made. In Finland, this issue has been solved by a) extending the scope of the Salvage Convention to wrecks not in danger, and b) by including a provision in the FMC that excludes salvage law from being applicable to wrecks covered by the national Antiquities Act. This Act defines a historical wreck as any wreck of a ship or other craft which can be presumed to have sunk at least 100 years ago. This yardstick is the same as in the UCH Convention. The Act does not imply that a wreck should be in danger in order to be protected.

Sweden and Denmark have also extended the scope of the Salvage Convention to wrecks not in danger. In addition, these countries have enacted similar national legislation excluding the application of salvage law to historical wrecks.[67] Latvian law requires a licence from the authorities before any activities concerning underwater cultural heritage can be undertaken, including salvage operations.[68]

Those BSCS which may not have extended the scope of the Salvage Convention to wrecks not in danger have passed national legislation that protects historical wrecks and evidently prevents salvage law from being applicable to such wrecks unless authorisation is obtained.

Which yardstick is applicable when defining the concept of historical wrecks is a question for each national legislature. The 100-year yardstick is used in some countries but different criteria are also used, e.g. in Sweden and Estonia. When

---

[66] Finnish Government Bill, note 33, at 8, 24.

[67] National law in Denmark excludes the collection of a salvage reward with regard to historical wrecks.

[68] Information received on 30 August 2016 from Professor Alla Pozdnakova.

Finnish law applies, historical wrecks cannot be the object of salvage without the authorisation of the authorities. The situation is presumably the same in the other BSCS.

6.9.3 Salvage of Wrecks Not Considered Historical

Under Finnish law wrecks, whether or not in danger and not considered historical, are covered by the Salvage Convention and the FMC. This is presumably the situation also in Sweden and Denmark. In the other BSCS, wrecks in danger but not classified as historical will, in principle, be covered by the Salvage Convention. If the scope of the Salvage Convention has not in these States been extended to vessels and wrecks not in danger, such vessels and wrecks are evidently covered by national salvage or similar legislation.

The leading principles of salvage law applicable to wrecks not classified as historical are the principles of encouragement and no cure-no pay; the criteria for fixing the salvage award are intended to encourage salvors to engage in salvage operations, but should the operation fail no award is granted.

Under the Salvage Convention and Finnish law, the owner of a wreck decides who is to have the right to commence a salvage operation of both the wreck and its cargo but is also entitled to prohibit the performance of such an operation. Anyone who salves a wreck in conflict with such a prohibition has no right to a reward provided, however, that the prohibition is express and reasonable. If not, the prohibition will have no effect and a reward can be collected.

An owner who has abandoned or can be considered to have abandoned their wreck will lose title to the wreck and the authority to conclude a salvage contract regarding the wreck. This may also open the door for third parties to claim title to the wreck, subject to the principles of appropriation. Such a new owner will then have the authority to decide on the salvage of the wreck.

Once a salvage contract is concluded, the salvor will have a right to fulfil their commitment and collect a reward, provided that the salvage operation has been successful. A salvor may be deprived of the whole or part of their reward due to fault or neglect on their part during the operation. A salvor who has prevented or minimised damage to the environment during a salvage operation may be entitled to special compensation for costs incurred.

It is held that if no salvor has been appointed, then the first salvor to commence a salvage operation may acquire a priority right, the right of the first salvor, to continue the operation. This right of priority entitles the salvor to rebuff other salvors. The owner of the wreck may still prohibit the first salvor from continuing the operation, provided that the prohibition is express and reasonable.

6.9.4   *Jurisdiction of National Courts*

The Salvage Convention, corresponding national law and national provisions on the protection of historical wrecks apply within the internal waters and the territorial sea of each of the BSCS. Almost all wrecks and historical wrecks registered by the Finnish authorities are located within Finnish internal waters and the territorial sea. This also applies to wrecks located in Swedish waters. Wrecks are also found within the territorial waters of the other BSCS. Approximately 60% of the total surface area of the Baltic Sea consists of territorial waters.[69]

Wrecks may also be found outside the territorial waters of the BSCS. Formally, they fall outside the application of national salvage law, the Salvage Convention and national provisions on protection. It has been held that potentially relevant jurisdictions with regard to salvage outside the territorial waters and territorial sea of a State could be the jurisdiction of the flag State of the wreck or the flag State of the vessel discovering or salving the wreck.[70] The State whose nationals have discovered or salved a wreck could perhaps also present corresponding claims. It has also been suggested that jurisdiction concerning ownership to a wreck or to property carried on board could be that of the flag State of the wreck, the law of the shipowner or the law of the nationality of the salvor.[71] This complicated issue will not be further considered in this paper. It can, however, be said that should the UCH Convention be generally accepted in the BSCS, the protection of historical wrecks would, to some extent, also cover areas outside the territorial waters of the BSCS.

In Lithuania, wrecks will be subject to the UCH Convention if they are located in territorial waters, in the exclusive economic zone or on the continental shelf. In practice, this convention will only concern wrecks older than 100 years, excluding them from the application of salvage law.

In Germany, Finland and Denmark wrecks, whether historical or non-historical, will also be governed by the WR Convention if they are located in territorial waters or in the exclusive economic zone. This convention does not have a direct impact on salvage law.

6.9.5   *Nature of Legal Regime*

If one wants to reflect on the nature of the legal regime applicable to salvage of wrecks prevailing in the BSCS, the following observations can be made. The Salvage Convention only concerns wrecks in danger, in principle, both historical and non-historical wrecks, which leaves the question of salvage of wrecks not in danger unregulated. This leaves room for the BSCS to fill this gap by introducing

---

[69] Varenius, note 8 at 21.

[70] Ringbom, note 17 at 27.

[71] Ringbom, note 17 at 27–28.

national or international law to govern the salvage of such wrecks, which has evidently been done in all the BSCS.

The distinction between historical and non-historical wrecks is of great importance when considering the scope of salvage law. In the BSCS, historical wrecks are protected and cannot usually be the object of salvage. Whether or not a historical wreck is in danger should be irrelevant. The age of a wreck is often decisive when considering protection, however, in at least Estonia, protection seems to be granted regardless of the age of a wreck, provided that the wreck can be held to be of historical interest. Only one of the BSCS has joined the UCH Convention, which sets the yardstick for protection at 100 years.

The legal regime in the BSCS governing salvage of wrecks and historical wrecks can with a slight exaggeration be described as an irregular patchwork quilt weaved around the Salvage Convention with elements from national law, the UCH Convention and from a general endeavour to protect underwater cultural heritage. Wrecks in danger and not considered historical are governed by the Salvage Convention. Every BSCS seems to have introduced their own provisions regarding the protection of historical wrecks. The salvage of wrecks not considered historical and not in danger are subject to national regulation. Wrecks not considered historical are fair game to salvors, though historical wrecks can usually not be the object of salvage. Uniformity might improve to some extent if all the BSCS were to join the UCH Convention. However, one should not forget that many of the BSCS have provisions on the protection of historical wrecks, which apparently offer better protection than the Convention, at least with regard to the age of the wreck required for the protection to enter into force.[72]

The WR Convention is today applied in Denmark, Finland and Germany while Sweden is preparing to join. It may only be a question of time before other BSCS will join. This would enhance uniformity in the field of wreck removal.

---

[72] The UCH Convention might accept national provisions that ensure better protection than those it has adopted (art. 6.1).

Chapter 7
Government Action Against Wrecks – A Finnish Perspective in Light of International Law

Markku Suksi

Abstract The chapter analyses the obligation of the public authorities and of the shipowner to remove a wreck. The legal setting is a fragmented one where several international agreements govern the matter, but few address it in detail. The focus of the chapter is on Finnish law - from the perspectives of pollution risk, traffic safety and of waste legislation - but an underlying theme is how the national rules are affected by and interact with provisions of international law.

Keywords Wreck removal · Intervention · Enforcement authority · Finland

7.1 Introduction

During the past few decades, some concrete measures taken in relation to ship-wrecks have been reported in the Finnish press. In 2001, a schooner with the name *Margona* went down at the waterfront in Helsinki, raising questions of what should be done. After the salvage and repair of the ship, it sank again in 2003 at a shipyard in Helsinki. The reports in the press made the point that the legislation in Finland is actually insufficient for dealing with wrecks, the implication being that under the law, it is unclear whether or not any authority, be it a public authority or the ship-owner, has any obligation to remove or make wrecks and their cargoes harmless. The article included a call for new legislation concerning wrecks. It seems that at least in the most recent case, an agreement was reached between the shipyard, the shipowner and the public authorities on the removal of the wreck.[1] The newspaper

---

[1] Hufvudstadsbladet, 14 December 2006, at 8. The ownership of the wreck was subsequently transferred to an association that owns the shipyard. According to the newspaper article, the owner-shipyard and the regional Trade, Traffic and Environment Centre of Nyland have concluded an agreement according to which the two parties pick up an equal share of the costs caused by the remains of *Margona*, estimated at 30000 euros.

M. Suksi (✉)
Department of Law, Åbo Akademi University, Turku, Finland
e-mail: msuksi@abo.fi

© Springer International Publishing AG, part of Springer Nature 2018
H. Ringbom (ed.), *Regulatory Gaps in Baltic Sea Governance*, MARE
Publication Series 18, https://doi.org/10.1007/978-3-319-75070-5_7

reader was nonetheless left with the impression that the public authorities were unable to do anything about wrecks because of a lack of competence.

However, Finnish law contains several provisions that deal with wrecks and cargoes that have gone down with the ship as well as with (hazardous) substances that remain on board the sunken ship. On the basis of the provisions, it is possible to take on such problems by considering the issue as detrimental to the landscape, as a blockage of the fairway likely to cause oil spills that pollute waters and vessels that can be categorized as litter as well as vessels which are in the process of being transformed into litter, *i.e.* vessels that technically speaking could still be repaired.[2] International law also contains rules on wrecks, but the international rules impact in different ways on national law. The result of this is a somewhat fragmented national regulation of what public authorities should and can do with respect to wrecks, which is not entirely satisfactory. Relatively recent developments at the level of international law through the Nairobi Convention of 2007 bring some clarity to the range of responses that public authorities have at their disposal. The issue here is thus the following: what is the obligation of the public authorities of Finland and of the shipowner to remove a wreck and how is this obligation affected by provisions of international law? The focus is here mainly on the national law of Finland from the point of view of risk of the pollution of waters, of traffic safety and of waste that wrecks may constitute.

7.2 Provisions of Relevant Legislation

7.2.1 Enactments of the Parliament of Finland

Section 25 of the Act on Oil Pollution Response 2009 (Finland)[3] establishes the following, which is based on the 1969 International Convention relating to Intervention on the High Seas in Cases of Oil Pollution Casualties[4]:

> If a ship sinks or runs aground in Finnish waters or in Finland's exclusive economic zone, becomes a party to a collision in said area, or is subject to a leakage or machine malfunction, or otherwise ends up in a state in which the risk of an oil spill or leakage of any other noxious substance is apparent, the Finnish Environment Institute may order the commencement of such rescue or other measures directed at the ship and its cargo that are considered necessary to preventing or limiting the pollution of water. Before taking such

---

[2] See M.-L. Kosola, S. Ruuska, J. Tuomainen, and M. Mäenpää *Romualukset ympäristöongelmana*. Raportti 19.3.2004 (Helsinki, Suomen ympäristökeskus, available at https://helda.helsinki.fi/bitstream/handle/10138/44744/Romualusselvitys_SYKE2004.pdf?sequence=1) 11. As pointed out in the report, at 15, the owner of a vessel, which is in a bad condition, is not necessarily interested in complying with requests to remove the vessel from the area, because he may have (unrealistic) plans for repairs to be made on the ship, plans that ultimately are not fulfilled. This article does not deal with the Antiquities Act 1963 (Finland), where s. 20 regulates the protection of wrecks that are more than 100 years old.

[3] Hereinafter: the OPR Act.

[4] UNTS 1975, vol. 970, No. I-14049.

measures, the Finnish Environment Institute must consult the Finnish Transport Safety Agency on the incident. Furthermore, the Finnish Environment Institute must consult the owner of the ship, the rescue company that has received the assignment, and the representatives of the insurers, if such consultations can be conducted without causing an unnecessary delay.

In order to prevent harmful consequences, the master of the ship that caused the water pollution or risk thereof must provide the authorities with any and all assistance required considering the circumstances.

The powers of the Finnish Environment Institute are in principle formulated as optional, because it *may* order such measures, and not as unconditional in such a way which would indicate that the Institute would have an obligation to take action. However, the context in which s. 25 of the OPR Act is placed indicates an obligation under ss. 1 and 5 of the OPR Act for the Environment Institute to take action, sustained by the provision concerning the responsibility for nature in s. 20 of the Constitution of Finland.[5] The obligation of Finnish authorities to take measures is also indicated by the Government Bill that led to the enactment of the Act on the Prevention of Pollution from Ships 1979 (Finland),[6] which seems to depart from the understanding that until the enactment of the 1979 Act, measures concerning vessels had, with the exception of damages caused by oil, depended on the ship-owner or the insurance company only, the implication being that from 1979 on, there is also a legal obligation on the part of the authorities of Finland to react.

Following this most immediate and swift form of reaction from the public authorities, another area, that of traffic safety, may also require a quick response to an incident that has resulted in the sinking of a ship. According to ch. 11 a, s. 5 of the *Maritime Act* 1994 (Finland)[7] there is an obligation on the part of the registered owner of the ship to take measures concerning the wreck after the Finnish Traffic Authority has, on the basis of ch. 11(a), s. 4, decided – on the basis of art. 6 of the 2007 Nairobi International Convention on the Removal of Wrecks – that the wreck constitutes a hazard,[8] thus the Traffic Authority shall, according to s. 5 of the

---

[5] Section 20 - Responsibility for the environment: "Nature and its biodiversity, the environment and the national heritage are the responsibility of everyone. The public authorities shall endeavour to guarantee for everyone the right to a healthy environment and for everyone the possibility to influence the decisions that concern their own living environment."

[6] Hereinafter: the PPS Act. See Government Bill 228/1978 at 6.

[7] As amended in 2016 and as entered into force on 27 January 2017 (see below, s. 4). After this, the provisions in s. 9 of the Decree of Marking of Fairways 1979 (Finland) on the removal of a wreck will be superseded by the new provisions in the Maritime Act, although the Decree itself might stay in effect in other respects. From the point of view of s. 2(3) of the Constitution of Finland, it is important that the public powers of the Finnish authorities concerning the removal of a wreck will no longer be established in this Decree, but in an Act of Parliament. From the point of view of the principle of *nullum crimen sine lege* in s. 8 of the Constitution, it is important that at least some of the penalties relevant in the context can now be based on the Maritime Act.

[8] 46 ILM 697. It is unusual that material domestic legislation in Finland makes such extensive reference to an international treaty, *i.e.* the references in the Maritime Act to the Nairobi Convention, although the Maritime Act itself contains a good number of references to other conventions of international maritime law.

Maritime Act, inform the registered owner of the ship as well as the competent public authority in the state of registration about this conclusion. After these procedures, the following shall take place on the basis of ch. 11(a), s. 5, sub-ss. 2–6, of the Maritime Act:

> The Traffic Authority shall establish a reasonable dead-line before which the registered owner shall remove the wreck and inform the owner about the dead-line in writing in the manner established in article 9.6 of the Nairobi Convention.
>
> If the wreck is considered to constitute a hazard according to section 4, the registered owner is under the duty to present evidence to the Traffic Authority on the existence of insurance or financial security referred to in section 7.
>
> In order to ensure safety at sea and to protect the marine environment, the Traffic Authority may, in accordance with article 9.4 of the Nairobi Convention, establish necessary conditions for the removal of the wreck and, in accordance with article 9.5. in that Convention, interfere in the removal. The Traffic Authority shall, before it decides on the matter, hear the opinions of the Traffic Safety Authority, the Finnish Environment Institute, and the Defence Forces.
>
> The Traffic Authority may remove a wreck at the expense of the registered owner in cases referred to in articles 9.7 and 9.8 in the Nairobi Convention when the site of the wreck is in Finland or within the economic zone of Finland.
>
> The Traffic Authority has the right to receive necessary executive assistance from the Defence Forces, the Customs and the emergency authorities for the fulfilment of its tasks according to this section.

The possible urgency in the implementation of this provision may in fact result in a situation where this provision is more or less simultaneously applicable with s. 25 of the OPR Act, above, although normally, the OPR Act should become activated before the above provisions in the Maritime Act. In addition, and beyond the short-term concerns expressed in the two provisions above, the Waste Act 2011 (Finland) contains in s. 72 the following prohibition of littering with a reference to vessels:

> No waste or discarded machine, device, vehicle, vessel or other object may be abandoned in the environment, and no substance may be emitted in a manner which may cause unclean conditions, disfigurement of the landscape, a decline in amenities, risk of injury to humans or animals, or any other comparable hazard or harm (*prohibition on littering*).

The enforcement of the Waste Act is mainly the responsibility of two different authorities, the regional Trade, Traffic and Environment Centre and the municipal environmental authority. Also, s. 18 (2) of the Environmental Protection Act 2014 (Finland) contains a prohibition of – apparently – the intentional sinking of vessels or scuttling:

> It is not permitted to dump within Finnish territorial waters or the economic zone waste or other substances with the intent of sinking or other intention of discarding from Finnish or foreign vessels, from vehicles on ice, from air carriers or from maritime technical units referred to in section 4(2) of the Act on the Protection of the Sea, and it is also not permitted to sink or abandon vessels, maritime technical units or air carriers by similar means outside of the economic zone. The same applies to dumping of substances into the sea from the shore with the intention of sinking or depositing them.

It should also be mentioned that in accordance with s. 7 of the Maritime Protection Act 1994 (Finland), the intentional sinking of ships in the sea, that is, the intentional creation of wrecks is prohibited in a similar manner by the Environmental Protection Act. Under s. 14 of the Maritime Protection Act, the Ministry of the Environment has the authority to exercise administrative force on the basis of the Administrative Fines Act 1990 (Finland) in order to enforce the Act.

Against the background of these provisions, it could be said that the legal order of Finland indeed contains provisions on the basis of which the public authorities can or even must react in situations where ships have sunk and that the authorities in fact have an obligation to take action, either directly in relation to the wreck and the cargo the ship was carrying, or indirectly through the owner of the ship, the operator of the ship, the salvage company or the insurer.[9] However, the material scopes of the provisions differ from each other.

7.2.2 Enactments Within the Legislative Competence of the Åland Islands

The above provisions apply in mainland Finland, but also in the Åland Islands to the extent the matter belongs to the legislative competence of the Åland Islands. Due to s. 18 of the Act on the Self-Government of the Åland Islands 1991 (Finland), the Legislative Assembly of the Åland Islands has the exclusive competence to enact Ålandic acts in the area of the environment for the jurisdiction of the Åland Islands. For this reason, a separate set of rules could apply to wrecks and their cargoes in the Åland Islands. This is also the case in relation to fairways for local sea traffic, while the Parliament of Finland has the law-making authority concerning maritime trade and fairways for such traffic. However, in the Åland Islands, the Legislative Assembly has not enacted any particular rules concerning wrecks comparable to the above provisions in the OPR Act or the Waste Act, except that there is a reference in the Ålandic Act on Measures Against Oil Pollution 1977 (Åland) to the Acts in mainland Finland that should also be followed in the Åland Islands and to the executive assistance that public authorities, that is, the authorities of the state, should give to the environmental authorities of the Åland Islands in case of an oil spill. The reason for this is the lack of legislative competence of the Åland Islands in certain respects concerning the regulation of fairways (but Åland has, on the other hand, legislative competence concerning water, the protection of the environment, waste as well as reception facilities on land for waste from ships, private boats for recre-

---

[9] See Kosola et al., note 2 at 6, for an account of practical instances in which it has been necessary to try to determine if and on what basis public authorities should act in respect of wrecks or ships that have been abandoned. On p. 7, the term wreck is defined as a vessel which because of an accident or for some other reason has sunk to the bottom of inland waters or the sea and of which a small part at the most is visible above the surface of the water.

ational use, and, as a consequence, also criminal law within these fields of legislation).

The Legislative Assembly of the Åland Islands adopted legislation in 1976 in this field, implying that the Government of the Åland Islands would have the authority to issue decisions concerning the necessary measures concerning a sunken ship, a ship that has run aground or become otherwise disabled, or concerning the cargo in order to prevent or limit oil damage. In its Opinion of 23 July 1976 to the President of Finland, the Supreme Court of Finland, under its authority to give opinions on the division of legislative competence between the Parliament of Finland and the Legislative Assembly of the Åland Islands, concluded that these provisions must be referred to the field of public maritime law. As a result of this, the Legislative Assembly had exceeded its competence and encroached into the sphere of the competence of the Parliament of Finland. Consequently, the President of Finland vetoed the Ålandic legislation.[10] Due to this division of competence in s. 27, para. 13, which places public maritime law within the sphere of competence of the Parliament of Finland, ch. 11(a) of the Maritime Act 1994 (Finland) that deals with the removal of hazardous wrecks and compensation for such removal (see s. 4 below) belongs to the legislative competence of the Parliament. For this reason, the relevant provisions of the Maritime Act concerning wrecks also apply within the jurisdiction of the Åland Islands, so, too, the OPR Act (see s. 3 below). However, as concerns the general norms dealing with environmental protection in maritime contexts[11] as well as concerning water and waste[12] the legislation of the Åland Islands applies and is implemented by the Government and municipalities of the Åland Islands.[13]

7.3   Risk of Pollution of Waters

Section 25(1) of the OPR Act and the previously effective s. 6 of the Act on the Prevention of Pollution from Ships (the PPS Act) are based on the 1969 International Convention relating to Intervention on the High Seas in Cases of Oil Pollution Casualties, or the so-called Intervention Convention.[14] Finland ratified the Convention on 6 September 1976 and it entered into force for Finland on 5 December 1976 (it had entered into force internationally on 6 May 1975). Finland incorporated the Convention into the national jurisdiction by Act (Statute number 810/1976) and Decree (Statute number 811/1976), which means that those provisions of the Convention that exist in the realm of ordinary legislation adopted by the Parliament are in effect in the internal legal order of Finland at the level of an Act of Parliament.

---

[10] See M. Suksi *Ålands konstitution* (Åbo Akademis förlag, Åbo, 2005) 241 (footnote 199).

[11] Ålandic Act on Countering Damage from Oil Spills 1977 (Åland).

[12] Ålandic Act on Environmental Protection 2008 (Åland).

[13] See Suksi, note 10 at 229–230.

[14] See Government Bill 248/2009 at 108, where reference is made to the fact that the current law as well as the previous national law are based on the Intervention Convention.

Arguably, this could be the case concerning art. 5 of the Convention, where a proportionality principle is established in order to determine which action is necessary on the part of the coastal state by taking into consideration the scope and likelihood of the threatening damage, the likely effectiveness of such measures, and the damage that may be caused by such measures. However, the article does not specify what measures the authorities could take, but leaves this for the states to decide. The latitude of the state is thus broad in choosing the suitable remedies.

When a ship has sunk, the Finnish Environment Institute may, on the basis of s. 25 of the OPR Act, order necessary rescue and other measures directed towards the vessel and its cargo, if there is a risk of the pollution of waters.[15] What sort of measures the law-maker envisions is, however, not clear on the basis of Government Bill 248/2009 for the OPR Act, because according to the bill, the provision corresponds, in the main, to the 1979 PPS Act.[16] This former Act envisioned, when being enacted on the basis of Government Bill 228/1978,[17] that such measures include the salvage of the ship, the unloading of the cargo and in extreme cases even the destruction of the vessel and its cargo by blowing it up. It is not clear on the basis of the Government Bill to the 1979 Act what positive consequences for the environment would result from the complete destruction of a wreck,[18] but the examples included in the Bill indicate that a broad range of different measures exist. However, what the measures could be is blurred by the fact that it is not Government Bill 248/2009, *i.e.* the bill to the current law, which lists them, but instead the bill to the former law.

The potentially unclear situation is nevertheless being somewhat structured by means of s. 23 of the OPR Act, which outlines the powers of the accident response authority. According to the provision, where necessary for preventing and responding

---

[15] Under s. 3, para. 5 of the OPR Act, oil damaged caused by a vessel means an instance or a series of instances with the same origin that is caused by a vessel and that leads to or can lead to an oil spill and that risks or can risk the state of water or the marine environment or the coastline or interests linked thereto and that require speedy action. According to J. Tuomainen *Vastuu saastuneesta ympäristöstä* (WSOY Lakitieto Helsinki, 2001) 276, it is well-motivated to depart from the point that the PPS Act is applicable to all vessels that have sunk or gone aground in Finnish waters, irrespective of when they sank. However, the PPS Act cannot be applied to vessels that sunk before the PPS Act entered into force. For such vessels, the duty to act held by public authorities is based on other provisions, such as the Waste Act 2011 and its predecessor, the Waste Treatment Act 1978 (Finland; the Act was repealed by the Waste Act 2011, but there is a possibility that the Waste Treatment Act 1978 is applied in respect of wrecks created before 1994. Hence the application of the PPS Act was relevant only in relation to new incidents, not to such pollution of the environment which already has taken place). In addition, at 281, Tuomainen points out that the PPS Act identified as primary actors the captain of the ship or the owner of the port, shipyard or other establishment, and only secondarily the public authorities charged with the duty to take measures against environmental threats.

[16] Government Bill 248/2009 at 107–108.

[17] Government Bill 228/1978 at 6.

[18] See N. Hooke *Modern Shipping Disasters 1963–1987*(Lloyd's of London Press, London, New York, Hamburg, Hong Kong, 1989) 481, dealing with the *Torrey Canyon* disaster in 18 March 1967, when the destruction of the vessel and its cargo was ordered to prevent an even greater environmental catastrophe: "The British Government gave orders that the *Torrey Canyon* be destroyed by bombing in the hope that all the estimated 40,000 tons of oil still remaining on board would be burnt off."

to oil spills or chemical spills from ships, and for limiting the consequences of such spills, the accident response authority shall be entitled to 1) temporarily commandeer any equipment and supplies suitable for accident prevention and response, any necessary communications and transport equipment, machines and tools, as well as premises and space needed for loading, unloading or temporary storage; 2) disembark and move about in another person's property; 3) order earth and water construction measures to be undertaken in another person's property; 4) limit waterborne traffic; and 5) take other measures necessary for preventing and responding to oil spills and chemical spills from ships. In spite of this listing, the contents of para. 5 still need to be understood on the basis of what was mentioned in the Government Bill to the 1979 PPS Act. This revoked Act was originally enacted for the implementation of the International Convention relating to Intervention on the High Seas in Cases of Oil Pollution Casualties, which is a direct result of the Torrey Canyon disaster.[19]

It is up to the Finnish Environment Institute to identify the measures, which in their most lenient form could even re-institute the ship for the regular use by its owner and which in their most extreme form could lead to the complete destruction of the ship and a total loss for the owner.[20] Measures in between these extremes include emptying possible oil or other containers and tanks that are on board the sunken ship, the isolation of the area of the incident by using oil restraints on the surface of the water and active cleaning measures for the surface of the water, shores, and even the bottom of the sea (if that is technically possible),[21] the cleaning of birds as well as the recovery of the ship and the cargo.[22] Section 11 of the OPR Act, dealing with executive assistance to the environmental authority from other public authorities, mentions that upon request and as far as possible, state authorities are obliged to provide executive assistance to the accident response authorities. The state authorities mentioned in s. 8(1) of the OPR Act are the Finnish Transport Safety Agency, the Finnish Defence Forces and the Border Guard. They shall participate in the prevention of and response to oil and chemical spills from ships, and according to s. 8(2), if they have detected or obtained knowledge of an oil or chemical spill from a ship, the authorities participating in the prevention and response operations must take urgent action to respond as necessitated by their level of preparedness, unless performing such tasks would significantly hinder their performance of any other important statutory task.[23] It is thus possible to say that the full

---

[19] For the *Torrey Canyon* disaster, see Hooke, note 18 at 480–481.

[20] In so far as the government agency and its staff is involved in a salvage operation, they should be entitled to salvage from the shipowner.

[21] Tuomainen, note 15 at 283.

[22] According to Tuomainen, note 15 at 286, immediate prevention and cleaning measures taken by public authorities do not require any additional environment permit.

[23] As concerns the defence forces, the mechanism of executive assistance from the defence forces to the Environment Institute established in s. 8 of the OPR Act is a special mechanism in relation to the general mechanisms of executive assistance established in the Act on Executive Assistance from the Defence Forces to the Police (781/1980). See also Government Bill 248/2009, in particular the motivations to s. 8 of the OPR Act.

force of the state can be put behind whatever measures are ordered by the Environment Institute.[24] The result of the actions of the Environment Institute can thus be the removal of the wreck and the cargo or the rendering harmless of the wreck or the cargo.[25]

The measures can in principle be ordered on the basis of an administrative decision of the Environment Institute.[26] The measures are exceptional also because they cannot, under s. 30, para. 5, of the OPR Act, be appealed against before any

---

[24] According to the Letter of Instruction of the Finnish Environment Institute of 15 December 2006 (SYKE-2002-P-126-044), a request of executive assistance presented by the Duty Officer of the Institute is valid and official even if made by phone and the Institute shall be liable for such a request. The Duty Officer shall confirm the request by a fax message or in some other way, although the validity of the request of executive assistance does not require this. In addition, the Letter makes the point that the Finnish Environment Institute shall cover the expenses caused by a request of executive assistance on the basis of a bill presented by the public authority that has given executive assistance. See also P. Wetterstein *Redarens miljöskadeansvar* (Åbo Akademis förlag, Åbo, 2004) 327, who concludes that the Act gives the public authorities broad powers to take measures. Tuomainen, note 15 at 276, concludes that the public authorities have been granted fairly broad powers to carry out damage control and other measures and that there exists no uncertainty about the right to act of public authorities. It should be taken into account that at least remotely, there could emerge a conflict with the protection of property under s. 15 of the Constitution. When balanced against s. 20 of the Constitution on responsibility to nature, especially in an emergency situation caused by a sunken ship, such a conflict should not arise.

[25] However, it should be stated at the outset that the Finnish Environment Institute has not used its powers even once under s. 6 to make a formal decision concerning measures (see below). It seems that the threat of such measures has the (unintended or perhaps even intended?) side-effect that the responsible party takes voluntary action. Nonetheless, the text below treats the matter as if such decisions could be made. A similar arrangement with a public authority furnished with broad discretionary powers is in place in most if not all EU member states. See also Directive 2002/59/EC of the European Parliament and of the Council of 27 June 2002 establishing a Community vessel traffic monitoring and information system and repealing Council Directive 93/75/EEC, art.s.19, concerning measures relating to incidents or accidents at sea: "1. In the event of incidents or accidents at sea as referred to in Article 17, Member States shall take all appropriate measures consistent with international law, where necessary to ensure the safety of shipping and of persons and to protect the marine and coastal environment. Annex IV sets out a non-exhaustive list of measures available to Member States pursuant to this Article. 2. The operator, the master of the ship and the owner of the dangerous or polluting goods carried on board must, in accordance with national and international law, cooperate fully with the competent national authorities, at the latter's request, with a view to minimizing the consequences of an incident or accident at sea. 3. The master of a ship to which the provisions of the ISM Code are applicable shall, in accordance with that Code, inform the company of any incident or accident, as referred to in Article 17(1), which occurs at sea. As soon as it has been informed of such a situation, the company must contact the competent coastal station and place itself at its disposal as necessary."

[26] According to Appendix 1, para. 9, of the Letter of Instruction of the Finnish Environment Institute of 16 February 2004 (SYKE-2002-P-126-044), the Duty Officer of the Finnish Environment Institute shall order, on behalf of the Institute, such salvage and other measures on the basis of s. 6 of the PPS Act (now s. 25 of the OPR Act) which are deemed necessary for the prevention or limitation of the pollution of waters. Before the measures are taken, negotiations referred to in the section shall be held. The same message is repeated in the Letter of Instruction of the Finnish Environment Institute of 15 December 2006 (SYKE-2002-P-126-044).

court of law,[27] something which further sustains the full force of the state behind the ordering of measures. This limitation of the right of appeal is very particular in light of s. 21 of the Constitution, which departs from the principle that there is a recourse to administrative courts in relation to administrative decisions by which a unilateral decision on the basis of law is passed concerning the right, benefit or duty of an individual and, in extension of individuals, of juridical persons. Because of this lack of remedies, the Environment Institute should not make decisions on ordering measures lightly, but should, even in potentially very stressful situations, where an environmental catastrophe may be unfolding, be careful in considering the exact measures that it decides to order. A prohibition of appeal may be justified, for instance, with reference to the speedy action that needs to be taken,[28] because administrative appeals would normally delay the implementation of a decision until a final court decision has been handed down.[29] The need for quick action in the environmental context would also, on basis of s. 31(2), *lit.* 4, of the Administration Act create a legal basis for not hearing the parties.[30]

Evidently, the Environment Institute has broad discretionary powers in its hands when it is put before a situation where a ship is sinking or has sunk.[31] Above, a broad range of potential measures was outlined, but this does not mean that the Environment Institute should, generally speaking, refrain from taking decisions, on the one hand, or resort to the blowing up of a wreck, on the other. Instead, s. 25(1) of the OPR Act makes reference to the concept of necessity: the measures ordered should be necessary for the prevention or limitation of the pollution of waters when there is a risk of such consequences.[32] The exercise of powers is not only limited to the water area of Finland, as identified in s. 2 of the OPR Act, but can, in addition, also be implemented with respect to incidents in the economic zone. It is debatable whether, for instance, the Air Force of Finland would heed a request from the Duty Officer of the Environment Institute to bomb a vessel so as to destroy it and the cargo completely,

---

[27] This prohibition of appeals is instituted in relation to the so-called ordinary appeals. The prohibition of ordinary appeals does not prevent so-called extraordinary appeals to the Supreme Administrative Court, which can result in the annulment of the decision, too, but the threshold of finding a violation of law is higher at the same time because filing an extraordinary appeal does not suspend the implementation of the decision.

[28] Government Bill 228/1978 at 6, underlines the fact that the measures to be undertaken should be extremely swift.

[29] Despite the prohibition of appeal, it could nevertheless be possible, if need be, to file an appeal with the Administrative Court, which could try the case directly on the basis of s. 21 of the Constitution if it comes to the conclusion that the prohibition of appeal in s. 33 of the Act is clearly in contravention with the Constitution in the manner established in and under s. 106 of the Constitution and thereafter try the legality of the decision.

[30] According to the provision, an administrative matter can be decided without hearing the party if the delay caused by the hearing may result in significant damage to the health of human beings, public safety or the environment.

[31] According to Tuomainen, note 15 at 276, the intention of the law-maker was to enact an Act which covers all potential pollution of waters caused by a vessel or situations that create such a risk.

[32] See also Wetterstein, note 24 at 328; Tuomainen, note 15 at 276.

because the effective provisions, in particular para. 5 of s. 23 of the OPR Act, are not explicit in outlining such drastic measures.

The reference to risk in s. 25(1) of the OPR Act seems to involve an idea of environmental impact assessment of some sort.[33] The idea in 1978 was certainly not to create an impact assessment procedure of the kind that is known on the basis of the Act on Environmental Impact Assessment 1994 (Finland). However, the Environment Institute is apparently expected to formulate a quick opinion of the relevance of the incident for the environment, that is, to perform an instant assessment of the environmental consequences that may follow from the incident.[34] No exacting analysis may be possible under such circumstances, but the assessment of the potential environmental consequences should involve at least information about the nature of the cargo and of the substances that remain on board the wreck. Under s. 25(2) of the OPR Act, the captain of the ship has the duty to assist the Environment Institute and other authorities of Finland with any information relevant to the matter, which means the captain should normally be able to provide the Institute with information on such substances on board the ship that are relevant for the assessment of the potential environmental consequences of the wreck and of the cargo.

The reference to considerations of necessity relates to the proportionality of the measures and underlines the fact that the measures should be dimensioned in a way which renders them effective for their purpose. They should neither be too lax nor too harsh. Instead, the measures should be sufficient to achieve the purpose, which is the prevention or limitation of the pollution of waters. However, under s. 19(1) of the OPR Act, the public authorities in charge of oil damage shall undertake all such urgent measures to prevent or limit damage, the costs of which are not in apparent disproportion to the economic and other values that are threatened.[35] The proportionality principle calls here for a differentiation between different situations so that the lowest effective level of substantive measures and the most cost effective level of measures is selected by the Environment Institute for the particular situation that has emerged.[36] Because accidents at sea are often very unique and singular without

---

[33] According to Tuomainen, note 15 at 276, the crucial question concerning the application of the 1979 PPS Act was what kind of a risk the vessel is creating and not, for instance, the fact of when or how the vessel has sunk.

[34] That an assessment should take place is also indicated by Appendix 1, para. 1, of the Letter of Instruction of the Finnish Environment Institute of 16 February 2004 (SYKE-2002-P-126-044), restating the powers of the Institute and the Letter of Instruction of the Finnish Environment Institute of 15 December 2006 (SYKE-2002-P-126-044) and informing other relevant authorities of the procedures concerning the operative measures.

[35] According to Tuomainen, note 15 at 282, the public authority has the power to take all sorts of measures to avert the pollution of waters, but the public authority should not take such measures which are unreasonably expensive in relation to the benefits. This is motivated because public authorities can recover costs from the liable party and from compensation funds only for well-founded expenses. See Tuomainen, note 15 at 286.

[36] However, according to Tuomainen, note 15 at 282, the proportionality considerations limit the measures ordered in relation to the subjective possibilities of the liable party, such as wealth or the size of the organisation. This, however, is not a regular concern from the point of view of administrative law, except when the administrative fines auxiliary to the performance of a primary obligation are determined.

there being much practice to refer to from before (that is, no scale of measurement exists), it would probably be reasonable to expect that the Environment Institute envisioned different accident scenarios involving ships so that a potential range of measures can be created in advance. In addition, the proportionality framework should contain the possibility of making a new decision with tougher measures if the first one is not effective enough. "Necessity" is also a consideration when the Environment Institute tries to make up its mind on whether or not it has to take measures. If the result of the consideration is negative, no formal decision recorded on paper is probably ever made (except if the request of a decision is formally initiated by an interested party with the right to file an application in the matter,[37] but even in such a situation, the Environment Institute should be able to explain why it did not consider it necessary to order measures). When a formal decision on the ordering of measures is made, reasons for the decisions must be given and the decision substantiated pursuant to s. 45 of the Administration Act 2003 (Finland).

There is another limitation of the discretion of the Environment Institute, which is that the decision to order measures must serve the specific purpose of averting the risk of the pollution of waters. The Environment Institute is acting *ultra vires* if it uses s. 25 for purposes other than in relation to a vessel in the water area of Finland that has sunk, is aground, has a leak or a machine malfunction or is otherwise in a comparable condition of incapacity which is not a "normal" failure of, for instance, machinery that can be remedied through regular towing or repairs.[38] The Institute cannot order measures, for instance, for the purpose of compelling a vessel (or the owner thereof) to pay such fees under the Act on Environmental Protection in Maritime Transport 2009 (Finland) for oil discharge or ship-generated waste or cargo residue charges it may have left unpaid when it last visited a Finnish port before it went down. In addition, the Environment Institute should observe the other limitations of discretion established at the level of the general principles of administrative law, that is, the principle of objectivity and the principle of equal treatment as they are mentioned in s. 6 of the Administration Act.

Although ordinary legal remedies are not available in a situation in which the Environment Institute orders measures, it can be said that the decision-making situation is not entirely free. The OPR Act creates a number of limitations for the discretion that the Environment Institute needs to take into account in its

---

[37] Such interested parties are identified in s. 186 of the Environmental Protection Act 2014 (Finland): 1) the party, 2) a registered association or foundation whose purpose is the promotion of environmental protection, the protection of health, nature conservation or pleasant living environments and within whose area of operations the environmental effects appear, 3) the municipality where the activities are located and any other municipality within which the environmental effects of the operations are felt, 4) the state oversight authority and the environmental protection authorities in the municipality where the activities are located and the municipalities in the area affected by them, and 5) an other authority with the task to watch over the public interest in the matter; 6) the Sami Assembly, if the environmental impacts appear in the Sami homestead area, and the village assembly of the Skolt Sami, if the environmental impacts appear within the Skolt Sami area

[38] See Government Bill 248/2009 108, for a general description of a situation of this kind.

decision-making, namely risk assessment and proportionality, but the legal purpose and suitability of the measures are also to be taken into account.

When making a decision on measures, the Environment Institute should, naturally, direct the decision towards somebody (that is, create the obligation to remove the wreck or the cargo), although the alternative remains that the Institute assumes the implementation of the measures itself with the support of executive assistance from other authorities (see above) or orders some other public authority to remove the wreck or the cargo. The addressee of the decision concerning the measures is informed by the Environment Institute about the existence of the obligation. The identification of the addressee (*i.e.*, identification of the shipowner or company for whom the obligation to remove the wreck is created)[39] of the decision concerning the measures should usually be relatively easy because the ship is normally registered with a national maritime authority in a country and carries an IMO identification number. However, in cases where the shipowner has been liquidated as a legal person because of bankruptcy or cannot be identified for other reasons, such a decision might, in fact, lack a "private" addressee.[40] In addition, under s. 25(2) of the OPR Act, the captain has the duty to assist the Environment Institute and other authorities of Finland with any information relevant to the matter. The owner of the cargo is not necessarily as easy to identify, but the captain should normally be able to provide the Institute with such information, too. In case the addressee of the decision concerning the measures is under Finnish jurisdiction, it should be fairly unproblematic to effectuate the decision. However, in cases where the addressee of the decision is not under Finnish jurisdiction but is, for instance, a company under the jurisdiction of a third state, it may be difficult to render the decision concerning the measures effective.[41] The measures ordered by the Environment Institute may in such cases have to be taken by the Institute and on the basis of executive assistance, and also the costs of those measures may then have to be borne by the Finnish state.[42]

Once the shipowner or the owner of the cargo has been identified, the Finnish Environment Institute can create an obligation for them to remove the wreck or the cargo by means of the decision concerning the ordering of the measures. The addressee has a *prima facie* obligation to fulfil the decision,[43] but in order to enhance

---

[39] See Wetterstein, note 24 at 328 f., who concludes that the addressee of the measures ordered by the authorities is generally the operator of the ship, although also other persons, such as the shipowner, can be considered an addressee.

[40] See also Tuomainen, note 15 at 278.

[41] According to Government Bill 228/1978 at 6, the right established in law to intervene may prove to be necessary, especially concerning foreign vessels. Hence, Finnish vessels were not considered a problem at the time of the enactment of the Act.

[42] According to the Letter of Instruction of the Finnish Environment Institute of 15 December 2006 (SYKE-2002-P-126-044), costs incurred because of executive assistance to other public authorities are covered by the Finnish Environment Institute.

[43] According to s. 43 of the Administration Act, an administrative decision should be delivered in written form. However, if a verbal decision is delivered, for instance, for reasons of expediency, then a written decision should be delivered as soon as possible.

the effect of the decision, the Environment Institute can, by separate decision on the basis of s. 28 on administrative enforcement, attach so-called administrative force to the primary decision concerning the measures. While the primary decision concerning the ordering of measures cannot be appealed, the separate decision to introduce administrative force can be appealed, as can the subsequent decision to implement the decision concerning the use of administrative force. The Environment Institute of Finland, shall, provided that this is clearly not necessary, sustain a prohibition or order of measures given on the basis of the OPR Act by an administrative fine or by the threat that a measure that has not been carried out is executed at the expense of the faulty party or that the activity is discontinued. Hence the primary decision involving the ordering of measures would, in most cases, involve the additional imposition of administrative consequences as established in the Act on Administrative Fines 1990 (Finland). It is important to distinguish the primary decision on the measures from the two supplementary decisions involving administrative force, namely the imposition of the fine or the threat and the execution of the fine or the threat. The function of the latter is to impose a pecuniary consequence for the addressee for the case that the primary decision is not implemented and to try to secure an economic return to the state of (at least a part of) those expenses that the implementation of the primary measures might cause in a situation where the addressee of the primary decision is not complying with the decision concerning the measures.[44]

The measures taken under the OPR Act are apparently meant to be immediate and swift in order to avert the realization of the risk to the environment. According to the Government Bill to the 1979 PPS Act, when a vessel in Finnish waters has run aground or been involved in some other maritime accident, it may, in order to avoid water pollution, be necessary to undertake particularly swift measures relating to the vessel.[45] This indicates that the PPS Act was – and the OPR Act is – intended to be implemented immediately after the incident leading up to the sinking of the ship or as soon as possible within a very short-term perspective. It may be possible to envision the implementation of the OPR Act from a mid-term or long-term perspective, too, but it appears that beyond the immediate or short-term perspective, a wreck and its cargo left in the water changes the legal regime from incident driven actions to actions dealing with waste under the Waste Act.[46]

---

[44] Section 35 of the OPR Act deals with the sharing of the economic burden following from measures taken under s. 25 of the OPR Act. According to the rule, the one who is liable for pollution of waters by action or omission relevant to the Act or by the emergence of a situation identified in s. 25, shall pay compensation. If the vessel, its cargo or other property on the vessel has, in a case that arises under s. 25, caused more damage than that which would normally be expected during salvage and if nobody can be expected to pay for the damage, the state shall compensate the excess amount under s. 38 of the OPR Act. See Tuomainen, note 15 at 282, who says that if the countermeasures prevent the damage, the expenses for them are the responsibility of the one who would be liable for the damage caused by the pollution of waters. For a graphic illustration of the mechanism on the use of administrative force, see Kosola *et al.*, note 2 at 23.

[45] Government Bill 228/1978 at 6.

[46] See also Tuomainen, note 15 at 282, who points out that the requirement of immediate action rules out measures of a less urgent nature.

Section 25(1) of the OPR Act establishes both an unconditional and a conditional duty to consult for the Environment Institute. The Institute has to consult those parties most likely interested in the fate of the wreck,[47] which are unconditionally the Finnish Transport Safety Agency, and conditionally – if such consultations can be conducted without causing an unnecessary delay – the owner of the ship, the rescue company that has received the assignment and the representatives of the insurers. The wording of the provision indicates that the decision concerning the ordering of the measures can be made before the consultations.[48] Evidently, the consultations can be used as a forum through which the Environment Institute assesses the willingness of the shipowner, the salvage company and the insurer to take measures so that the Institute can determine the extent of administrative force that should be attached through a separate decision to the decision concerning the measures to be taken. Because the duty to consult the shipowner, the salvage company and the insurance company may become operative only after the decision on measures has been made, the duty is not such a legal basis upon which the decision concerning the ordering of the measures could be revoked on the basis of extraordinary administrative remedies. The duty to consult is, however, not unconditional (except for the Finnish Transport Safety Agency), but is mitigated by a consideration of the time perspective. In a situation where negotiations with the interested parties could cause a detrimental delay to the implementation of the measures ordered by the Environment Institute, the measures can also be commenced before the consultations. In addition, the duty to consult does not mean that an agreement should be achieved between the parties. It is probably, formally speaking, enough that such consultations are arranged between the interested parties, but they do not have to result in an agreement concerning the implementation of the measures by the different parties.

Against the background of the OPR Act, it is submitted that the Environment Institute has at least an obligation to consider, in conjunction with an incident, whether the incident constitutes such a risk for the environment that could warrant measures ordered by the Institute.[49] This is so in spite of the fact that after such consideration, the Institute has a great measure of discretion in deciding whether or

---

[47] As indicated by Appendix 1, para. 9, of the Letter of Instruction of the Finnish Environment Institute of 16 February 2004 (SYKE-2002-P-126-044) and the Letter of Instruction of the Finnish Environment Institute of 15 December 2006 (SYKE-2002-P-126-044), the Duty Officer of the Finnish Environment Institute is the civil servant with the full powers to represent the Institute in such negotiations.

[48] This also seems to be the procedure recorded in Appendix 1, para. 9, of the Letter of Instruction of the Finnish Environment Institute of 16 February 2004 (SYKE-2002-P-126-044), according to which the Duty Officer of the Environment Institute shall carry out negotiations before the commencement of the measures.

[49] According to Appendix 1, para. 1, of the Letter of Instruction of the Finnish Environment Institute of 16 February 2004 (SYKE-2002-P-126-044), the Duty Officer shall, after having received an incident report or other information about an incident, assess the incident in order to determine, whether an "OPR-situation" is at hand and thereafter determine the scope and seriousness of the incident.

not to order measures and what sort of measures it orders. After all, the provision in s. 25(1) of the OPR Act uses the term "may". If the Institute fails to engage in such a consideration and the incident results in damage to the environment, three possible reactions to the inaction of the Institute could potentially emerge, namely disciplinary measures under the Civil Servants Act (750/1994) and even severance from office on the basis of the same Act, criminal responsibility under s. 40:9 or 40:10 of the Penal Code 1889 (Finland) for failure to fulfil the official duties, and the tort liability of the state as the principal actor under s. 3:2 of the Damages Act 1974 (Finland) for damage caused in conjunction with the exercise of public authority (which, in a situation of this kind, would imply negligence). The criminal and tort liability of the leader of the prevention measures is mentioned in s. 10(2) of the OPR Act.

The main issue in this context is, of course, how can it be established that the consideration of an incident has taken place. Under the internal instructions of the Environment Institute, the Duty Officer of the Institute shall maintain a diary of his or her actions.[50] If information about an incident has been delivered to the Environment Institute and no action has been taken, the Environment Institute and the Duty Officer would be at fault unless the diary contains an entry that establishes a consideration of the matter and reasons for inaction. The diary, again, is a document that should arguably be considered a public document under the Act on the Publicity of the Activities of Public Authorities 1999 (Finland), which means that it should be possible for, *e.g.* other public authorities and the media to find out information about an incident and about the consideration that has gone into the case. The Environment Institute has thus a primary obligation to react and, on the basis of the determination of the Institute, a secondary obligation can be created for the operator of the ship.

In practice, the Finnish Environment Institute has not made any decisions on the basis of s. 25(1) of the OPR Act (or the former s. 6 of the PPS Act). Ordering measures on the basis of the provision constitutes a means that has not been necessary to use so far. Consequently, there exists no attempt to try such decisions in courts of law.[51] The Institute has so far utilized its discretion (under the term "can" in s. 6 of the revoked PPS Act and) under the term "may" on the basis of s. 25(1) of the OPR Act on the basis of the interpretation that s. 25(1) of the OPR Act does not presuppose a formal decision. Hence, no tangible examples of practical application of s. 25(1) exist in the form of decisions of administrative authorities or courts of law. The government has so far been very careful in avoiding the formal application of the rule. Instead, if concrete situations have at all existed, it seems that the option of action in the hands of the Environment Institute to implement close to "dictatorial" powers has been sufficient to persuade the shipowner and the captain of the vessel

---

[50] Letter of Instruction of the Finnish Environment Institute of 15 December 2006 (SYKE-2002-P-126-044).

[51] Telephone interview with Mr. Kalervo Jolma, team-leader of the environmental emergency response unit, on 14 May 2007 and e-mail message on 16 May 2007.

to co-operate on the matter and establish working-relationships (the captain of a vessel anyway has an interest in avoiding damage and a duty to act in a manner that damage is avoided).

There is a specific Duty Officer at the Finnish Environment Institute in constant stand-by during every hour of the day seven days per week, who, in practice, is the sole implementing agent of s. 25(1) of the OPR Act. This means that there is an alarm system into which information about an incident is fed and which is supposed to lead to the recognition of a problem that may require action under s. 25(1) of the OPR Act.[52] In practice, the principle is that the wishes and requests of the Duty Officer are fulfilled even without a formal decision-making on the basis of s. 25(1). The Duty Officer is in charge of the salvage operations and is directly authorized to activate 14 ships and two airplanes.[53] In addition, for limiting damage caused by oil, the Duty Officer can activate the necessary resources for clean-up work on the basis of the OPR Act.

According to the Finnish Environment Institute (fi: SYKE): "when an incident has happened, the SYKE's officer on duty:

- represents the Finnish Environment Institute and acts on SYKE authority until otherwise prescribed;
- calls for executive assistance from other authorities;
- dispatches recovery vessels and equipment;
- initiates and co-ordinates and is in charge of recovery efforts and appoints the leader for the response operation;
- acquires other necessary materials or staff;
- is responsible for informing other organisations about the incident according to international agreements (e.g. HELCOM) and informing in general;
- requests international assistance;
- initiates required investigations."[54]

"When an oil spill is observed at open sea, the report may be given to the nearest Maritime Rescue Coordination Centre (MRCC/MRSC) via coastal radio, pilot or

---

[52] See Letter of Instruction of the Finnish Environment Institute of 16 February 2004 (SYKE-2002-P-126-044) and Letter of Instruction of the Finnish Environment Institute of 15 December 2006 (SYKE-2002-P-126-044) informing other relevant authorities of the procedures concerning the operative measures. In the latter letter, the Institute requests relevant authorities and institutions that their subordinate units, such as duty stations (e.g., pilot stations and VTS centres) and ships immediately inform the closest sea rescue centre or alarm centre and, when possible, also the Finnish Environment Centre, *i.e.* its Duty Officer on the environmental damages of, *inter alia*, all OPR-related incidents or risks thereof. For a graphic illustration of the alarm system, see http://www.ymparisto.fi/download.asp?contentid=28807&lan=en

[53] Telephone interview with Mr. Kalervo Jolma, team-leader of the environmental emergency response unit, on 14 May 2007.

[54] Available at http://www.ymparisto.fi/en-US/Waters/Environmental_emergency_response_in_Finland/SYKEs_officer_on_duty

Table 7.1 Incidents between 2000 and 2015 involving vessels and oil spills

Year	Vessel accidents that required action against oil spill	Other accidents with vessels	Other oil spills on the sea
2000	6	12	89
2001	6	14	76
2002	3	10	95
2003	2	22	94
2004	1	10	66
2005	3	17	49
2006	4	18	65
2007	1	23	60
2008	3	22	76
2009	3	12	56
2010	1	20	37
2011	–	27	46
2012	2	13	29
2013	3	13	29
2014	–	24	51
2015	–	64	55

Coast Guard stations; they will further inform SYKE's officer on duty."[55] From this report, which is sent to the Duty Officer, machinery is set in motion that implies the activation of a number of different public authorities and other contact points.[56]

The following incidents involving vessels and oil spills have been reported to the Finnish Environment Institute between 2000 and 2015 (Table 7.1).[57]

However, the powers under s. 25(1) of the OPR Act (or its predecessor the PPS Act) were not activated in one single instance.

7.4 Considerations of Traffic Safety

Section 5 in ch. 11(a) of the *Maritime Act* 1994 (Finland), as amended in 2016 by Act 860/2016, is interesting because it creates a mechanism for the removal of wrecks that is in unusually close material proximity with provisions in the 2007 Nairobi International Convention on the Removal of Wrecks,[58] which entered into force internationally on 14 April 2015 and was ratified by Finland on 14 October

---

[55] Available at http://www.ymparisto.fi/en-US/Waters/Environmental_emergency_response_in_Finland/SYKEs_officer_on_duty

[56] The operational environment of the Duty officer is available at http://www.ymparisto.fi/en-US/Waters/Environmental_emergency_response_in_Finland/SYKEs_officer_on_duty

[57] See Öljyvahinko- ja erityistilannepäivystyksen tapahtumat 1985–2015, available at http://www.ymparisto.fi/fi-FI/Vesi/Oljy_ja_kemikaalivahinkojen_torjunta/Ymparistovahinkopaivystys

[58] Government Bill 96/2016.

2016. Finland incorporated the Convention into the national jurisdiction by provisions in the Maritime Act and the *Government Decree on the Bringing into Force of the International Convention on Removal of Wrecks and on the Entering into Force of Those Provisions in the Convention that Belong to the Sphere of Legislation* 2017 (Finland), which means that those provisions of the Convention that exist in the realm of ordinary legislation adopted by Parliament are in effect in the internal legal order of Finland at the level of an Act of Parliament as of 27 January 2017. Arguably, this could be the case concerning art. 9 of the Convention, where a range of actions on the part of the coastal state is established for the removal of a wreck. The material law in Finland, that is, provisions in ch. 11(a) of the Maritime Act, makes ample reference to the Nairobi Convention, linking measures taken by Finnish authorities closely with the relevant provisions of the Nairobi Convention.

Section 5 of ch. 11(a) of the Maritime Act contains a clear obligation to remove such a wreck within Finnish waters or the Finnish economic zone that may endanger the traffic of vessels. The material duty created by the Maritime Act entails the removal of the wreck, but does not extend the duty to any sunken object, ranging from the entire vessel through to a part of the vessel, to an object that has sunk, such as part of the cargo (e.g., a container), which s. 9 in the Decree of Marking of Fairways 1979 (Finland) does. In this respect, the provisions of the Maritime Act seem apply to less "objects" than the Decree, which might be problematic if some *lacunae* remained outside of the definition of the concept of wreck in art. 1(4) of the Nairobi Convention[59] and the requirement that the wreck should be a hazard. However, a wreck referred to in the Maritime Act should probably be understood in the broad meaning of the term as established by art. 1(4) of the Nairobi Convention. There is also a provision in s. 1(4) of ch. 11(a) of the Maritime Act that purports to distribute governmental action between the Maritime Act and the OPR Act: the provisions of ch. 11(a) of the Maritime Act shall not be applied in a situation where measures against a risk of oil and chemical damage caused by a wreck are undertaken on the basis of the OPR Act, which actually operates as a distribution of jurisdiction between the Traffic Authority and the Finnish Environment Institute.

The term "removal of wreck" in s. 5 is interpreted within art. 1.7 of the Nairobi Convention to include every prevention, alleviation or eradication of hazard that a wreck constitutes. The measures prescribed encompass only wrecks that constitute

---

[59] However, as established in art. 1(4), paras. b and c, cargo is also defined as a wreck: "(b) any part of a sunken or stranded ship, including any object that is or has been on board such a ship; or (c) any object that is lost at sea from a ship and that is stranded, sunken or adrift at sea." The definition of a wreck in the Nairobi Convention is therefore very broad and might result in a situation where no *lacunae* remains. At the moment of writing this article, the Decree remains in force, but the new provisions of the Maritime Act have the effect of replacing the relevant provisions of the Decree, even though the Decree itself would not be repealed (although such a repeal is probably forthcoming). For an analysis of the Decree before the new provisions of the Maritime Act, see M. Suksi, "Government and Wrecks – On the Obligation of Public Authorities and/or of the Owner to Remove or Make Harmless Wrecks and their Cargoes", in H. Rak and P. Wetterstein (eds) *Shipwrecks in International and National Law* (Institute of Maritime and Commercial Law of Åbo Akademi University, Åbo, 2008) 146–148.

under s. 4 of ch. 11(a) a hazard for safety at sea or for the marine environment.[60] Because the fairway in which the incident has taken place is likely to be used by other seafarers, there is a sense of urgency in the removal measures.[61] However, in so far as the wreck is situated outside fairways in a place where more regular water traffic does not occur, the removal is not necessarily a matter of urgency, because there is not necessarily any hazard involved, although the duty to remove in principle remains, because water traffic can take place in all sorts of waters. Nonetheless, in comparison with the OPR Act, the geographical extent of the duty is probably more limited, because it concerns waters that are used by seafarers (and thus where a wreck can cause a hazard), whereas the OPR Act may result in measures over a vast geographical area.

The bearer of the duty to remove under the Maritime Act is the owner of the ship, who can do it by himself or contract a salvage company or some other salvor for the task of removing a hazardous wreck.[62] According to s. 4, the Traffic Authority shall inform the owner of the hazardous wreck and set, on a case-by-case basis and by considering the nature of the hazard caused by the wreck, a reasonable deadline, communicated in writing to the owner, within which the owner shall remove the wreck. The Traffic Authority shall also inform the owner of the nature of the wreck and about the measures required to remove the wreck. In addition, the owner of the ship shall be notified of the fact that if the owner does not remove the wreck, the Traffic Authority may remove the wreck at the expense of the owner.[63] The Act does not spell out the manner in which the state could claim the expenses from the owner, but the Act departs from the position that the ship is insured in the manner envisioned by art. 9(3) of the Nairobi Convention, which would seem to indicate that compensation for removal should be forthcoming in all situations.

Before the removal operation commences, the Traffic Authority may establish such conditions for the removal of the wreck that are necessary for traffic safety at sea and for the marine environment.[64] The conditions can be issued to the owner of the ship or the salvage company engaged by the owner and they may be issued when such conditions are necessary for ensuring safety at sea or for the protection of the marine environment. For the same reasons, the Traffic Authority may interfere in the removal of the wreck after the removal has commenced. The Traffic Authority should, when informing the owner of the deadline for removal, also notify the owner

---

[60] Government Bill 96/2016 at 29

[61] See Wetterstein, note 24 at 325, and Kosola *et al.*, note 2 at 18. Under s. 13:4(1) of the Water Act (587/2011), the owner of the ship is liable for compensation of losses caused by public traffic, if the loss has been caused on purpose or is due to recklessness, unless otherwise provided by the Maritime Act.

[62] It seems that the shipowner is not in a position to transfer his duties to a third party in order to free himself from the obligation. See also Wetterstein, note 24 at 327. In addition, see Wetterstein, note 24 at 326, according to whom it is natural to hold the operator of the ship as the duty-bearer for the removal of both the ship and such cargo that has sunk. However, according to the new law of 2016, the duty is instead on the owner of the ship.

[63] Government Bill 96/2016 at 30.

[64] Ibid.

that the Traffic Authority intends to interfere in the removal if the hazard is threatening to become particularly serious.[65] Before decisions by the Traffic Authority concerning the introduction of conditions or interference in the removal operation, the Traffic Authority shall hear the Traffic Safety Authority on issues related to safety at sea, the Finnish Environment Institute on issues related to the protection of the marine environment, and the Defence Forces on issues related to the infrastructure of the Defence Forces in the marine areas in question.

As mentioned in the Government Bill 96/2016, the Traffic Authority may, under s. 5, sub-s. 5, of ch. 11(a) of the Maritime Act, remove a wreck at the expense of the registered owner of the ship in cases mentioned in arts. 9.7 and 9.8. of the Nairobi Convention when the wreck is situated in Finland or within the Finnish economic zone. The Traffic Authority may thus remove the wreck when the registered owner is not doing so within the deadline established by the Traffic Authority or when the owner cannot be reached or in situations where immediate measures are required. In situations where immediate measures are required and where the wreck is within the Finnish economic zone, the Traffic Authority must inform the registered owner of the ship and, in case the ship is foreign, also the state of registration of the ship. The relevant provisions of the Nairobi Convention establish a number of administrative principles for the removal operation according to which a wreck shall always be removed by using the best available measures that are most practical and efficient and by considering safety at sea and the protection of the marine environment. The Traffic Authority shall, when considering the measures by which to prevent, alleviate or eradicate a hazard, take into account the general principles of art. 2 of the Nairobi Convention, under which the measures shall be proportionate to the hazard and not exceed what could be reasonable to remove a wreck that constitutes a hazard. The measures shall also cease as soon as the wreck has been removed. In addition, the measures shall not unnecessarily limit the rights and interests of other states or of physical or juridical persons.[66] These administrative principles should guide the decision-making authority, thereby circumscribing the discretion at the instance of decision-making. However, when enacted, the measures are intended to be swift, because under s. 8 of ch. 11(a) of the Maritime Act, the decisions of the Traffic Authority may be implemented immediately, even if they have not yet acquired finality, unless the decision itself or the appropriate first instance court with maritime jurisdiction determines otherwise.

The Traffic Authority has the right to receive the necessary executive assistance from the Defence Forces, Customs and from the emergency services for its tasks under ch. 11(a), s. 5, of the Maritime Act. In addition, the Traffic Authority has the possibility to receive executive assistance from the Border Guard on the basis of s. 77 of the Act on Border Guard 2005 (Finland).[67] It seems that also here, the full force of the state can be placed behind the measures that are deemed necessary to remove the wreck.

---

[65] Ibid.

[66] Ibid.

[67] See also Government Bill 96/2016 at 31.

7.5   Wrecks as Waste

Section 72 of the Waste Act 2011 (Finland) with its prohibition of littering contains
a reference to vessels that indicates vessels must not be abandoned in the environment in a way that turns them from a being a useful object into litter.[68] A question
significant after the above review of provisions related to the prevention of the pollution of waters and to traffic safety is the point of time when a wrecked ship and/or
its cargo become waste in such a way that it can be regarded as litter. According to
Government Bill 77/1993 for the 1994 Waste Act, the essential criterion is that the
object (in this case a vessel or an object that has been transported by the vessel) has
been taken out of use in such a way that a detrimental consequence of the kind that
the Act identifies then emerges.[69] In the last instance, the public authorities are
called upon to determine whether or not the vessel can be considered litter or not.
The owner of the vessel may, however, be of another opinion concerning the condition of the ship than the owner of the water area or the public authority that deals
with the issue, in most cases the municipality and its environmental board.[70]
According to Kosola *et al.*, a vessel is transformed into litter at the point when the
vessel is taken entirely out of duty and use. According to them, such a vessel can be
considered as taken out of duty and use when it has been declared irreparable.[71]
This, however, is debatable, and in practice, there may exist a grey area between the
declaration of irreparability, which is particularly relevant from the point of view of
the insurer, and the emergence of the vessel as litter. After a declaration of

---

[68] According to J. Tuomainen *Vastuu pilaantuneen ympäristön puhdistamisesta*(Suomalainen
lakimiesyhdistys, Helsinki, 2006) 184, the provision in the Waste Act concerning littering can be
applied in the case of movable waste on the bottom of a water area, such as containers of chemical
substances and shipwrecks. Tuomainen also makes the point that the prohibition of littering is not
applied regarding the pollution of the water, for which provisions exist in the now repealed
Environmental Protection Act (86/2000). Tuomainen also concludes, at 214, that the essential
point for considering an object as litter is that the object has been taken out of use in such a way
that a detrimental consequence, as defined by the provision in s. 3 of the Waste Act, is created, *i.e.*
it is a risk to health, has a negative consequence on the environment, *e.g.* it degrades the landscape,
or diminishes its enjoyability, or there a comparable risk or detriment.

[69] See Government Bill 77/1993 at 27. The Government Bill refers to current practice as an indica-
tor of why the prohibition of littering would also be relevant for a large object, such as a vessel,
thus this reference is evidence for the existence of a legal interpretation of large vessels as litter
even before the enactment of the Waste Act. Also, Tuomainen, note 68 at 215, concludes that a
vessel taken out of use can be regarded as a breach of the prohibition of littering, if it causes untidi-
ness, becomes an eyesore in the environment or any other detriment. See also Tuomainen, note 15
at 280, in which the point is made that it is possible to interpret the prohibition of littering so that
it also applies to a sunken ship from which a chemical substance is leaking and causing danger to
health or the environment.

[70] See Kosola *et al.*, note 2 at 11.

[71] Under s. 1:7(1) of the Maritime Code (674/1994), a vessel that has been damaged shall, during
inspection, be declared to be irreparable when reparation is not possible or if the vessel can not be
transferred to a place where the repair works should take place or if it is not worth repairing the
vessel. See also Kosola *et al.*, note 2 at 17.

irreparability, there still may exist such property values in the vessel (such as scrap metals), which could lead to the vessel having other end results than becoming litter.

Under s. 73 of the Waste Act, the person responsible for litter has the obligation to remove the object or substance from the environment and otherwise clean up the littered area. The litterer has thus the *primary* duty to clean the area. In principle, this means that any of the shipowner, the operator of the ship or the captain of the ship could be the possible addressee of the prohibition of littering and thus be responsible for cleaning up the littered area. The formulation of the prohibition of littering nonetheless indicates that the litterer that is primarily liable for the cleaning efforts is a subject which has been in active control of the object that is classified as litter.[72] Concerning wrecks and cargoes that have gone down with vessels, the shipowner is probably not an active party liable for littering, instead the operator of the ship or the captain is likely to be.[73]

In case it is not possible to identify or reach the litterer or if the litterer fails to comply with his cleaning duty,[74] it is possible that under s. 74(1) of the Waste Act, the holder of the littered area and, in the area of, for instance, a public outdoor leisure or motorsledging route, the keeper of the route can be compelled to assume responsibility for cleaning the area.[75] The holder of the littered area may thus become liable, in the second instance, for the cleaning, which, in the case of water areas, means that there may be a multitude of holders, such as the formal owner of the water area (private individual or legal person owning real property and water, an association of owners of real property, the municipality and the state) or the holder of the lease of a water area. This means that it should normally be fully possible to identify the bearer of the *secondary* duty to clean, although this may simultaneously mean that the cleaning duty is atomized to a great extent due to the great number of owners and holders of water areas. There is a *tertiary* bearer of the cleaning duty, which under s. 74(2) of the Waste Act is the municipality in which the litter is to be found, provided that the secondary holder of the cleaning duty has failed to comply with the cleaning duty or if the duty of the secondary instance to clean must be

---

[72] Although immediate prevention and cleaning measures may have been taken under the OPR Act under the authority of the Finnish Environment Institute and without any specific environment permit, secondary measures may require an environment permit on the basis of Environment Protection Act. This is also true if the environmental authorities believe that incomplete and insufficient measures have been taken under the OPR Act and that new cleaning measures must be ordered. See Tuomainen, note 15 at 286.

[73] See also Kosola *et al.*, note 2 at 11.

[74] As concluded in Tuomainen, note 15 at 277, the owner of a sunken ship or an insurance company which has taken on the rights of the owner may not necessarily have a great interest in salvage operations to which costs for environmental liabilities are attached. In addition, Tuomainen points out that although the owner can, under private law, relieve itself from ownership of the vessel, such transfer of ownership can not entail the transfer of obligations under public law.

[75] It seems that routes comparable to public outdoor leisure routes could also be identified on water. See also Kosola *et al.*, note 2 at 16.

considered excessive.[76] Hence it should be possible at all times to find an answer in legislation to the question of who is responsible for removing the vessel.[77]

In practice, it has been, from time to time, difficult to determine the primary and secondary instance that would be responsible for the wreck. The owner has perhaps been insolvent, and it is also sometimes impossible to sort out who the owner really is, because transfers of ownership have perhaps not been communicated to the ship register. Therefore, issues concerning wrecks have often been dealt with by municipalities at the local government level.[78] In most cases, the Waste Act has been applied to the wreck and in such situations the party that has caused the environmental hazard or consequence or the owner of the area has been charged with the duty to remove the wreck. The environmental authority of the municipality has laid the responsibility on the municipality itself in several cases. The municipality has normally assumed the costs in situations where the wreck is in the area of a harbour owned by the municipality or when the wreck is situated in the municipality.[79]

Under s. 75 of the Waste Act, the municipal environment committee is the primary public authority that can make decisions with a reference to the prohibition of littering. An environment committee can order a litterer to clean a littered area, and if the litterer is not identified, the committee can order another party – obliged under the law to clean litter –to do so.[80] This means that a wreck or sunken cargo that is no more actively the object of actions on the basis of the OPR Act or the Maritime Act and, in so far it has been abandoned, not anymore the object of actions by the shipowner, salvage company or insurer, can be the material object of an administrative decision by a municipal environment committee.[81] In such a decision, the environment committee must be able to identify an addressee for the decision, which is either the primary party that has caused the litter (operator of the ship or the captain) or another party that may, under the law, have a secondary obligation to clean the litter.[82] This is so also because of s. 3 of the *Environmental Protection Act* 2014

---

[76] Kosola *et al.*, note 2 at 16. According to Tuomainen, note 68 at 184, 191, in the year 1993 the Supreme Administrative Court of Finland defined, on the basis of the 1978 Waste Treatment Act (673/1978), a municipality as the duty-bearer obliged to take away the lower part of a vessel that had sunk in the 1960s close to a beach area and, in 1998, the decision of the Court placed a duty on the municipality to remove from the water area a vessel which had caused the environment to become unsightly. See also Tuomainen note 15 at 279, and Kosola *et al.*, note 2 at 16.

[77] Kosola *et al.*, note 2 at 11, 26.

[78] Kosola *et al.*, note 2 at 10.

[79] Kosola *et al.*, note 2 at 11.

[80] According to Tuomainen, note 68 at 215, the Supreme Administrative Court ordered in 1997 the taking away of a vessel. Tuomainen draws the conclusion that in line with an established line of interpretation by the Supreme Administrative Court, it is possible to order the taking away of a vessel on the basis of the prohibition of littering.

[81] Concerning vehicles, Tuomainen, note 68 at 215, presents the example that if the owner of the vehicles declares that it has the intention to use the vehicles at a future point of time, the public authority shall inquire into whether the objects have been taken out of use. This interpretation would bring the time of application of the Waste Act closer to the incident.

[82] As concluded by Wetterstein, note 24 at 331, the Waste Act is accorded with special relevance regarding the taking away of wrecks and objects on board that have been abandoned, even when they do not constitute a threat to sea traffic or constitute a risk of pollution of waters.

(Finland) (as was the case in s. 2 of the revoked Environmental Protection Act 2000): the Environmental Protection Act does not apply to activities prescribed upon in the OPR Act. This would indicate that the prevention of environmental pollution under the OPR Act is primary in relation to actions taken under other legislation.

As a practical matter, public authorities would first try to use so-called "soft" measures when public authorities start to deal with the matter. The public authorities will therefore first try to explain the illegality of the situation to the responsible party and to set out the measures that are potentially needed. On the basis of this first contact, the public authority can step up its involvement in the matter by presenting the responsible party with a request for a written report concerning the reasons for the incident, the measures that have been taken so far and the measures that will be taken in the future. Unless the responsible party removes the vessel in the manner that the public authorities expect, the competent public authority can give a written request in which the responsible party is obliged to take measures for restoring a legally acceptable situation during a certain time-frame. The measures and the timetable for their implementation shall be defined against the background of considerations of the possible consequences of the offences that may be involved, possibilities to restore the legally acceptable situation and the efficiency of the measures. The responsible party shall have the right to reply.[83]

Once the regional Trade, Traffic and Environment Centre or the municipal environment committee have made decisions under the Waste Act, it is possible for the addressee of the decision to appeal the decision under s. 137 of the Waste Act. The appeals procedure follows the provisions of the Administrative Judicial Procedure Act 1996 (Finland) and the appeals are filed at the provincial administrative court, with the possibility to appeal the decision of the provincial administrative court at the Supreme Administrative Court. However, in so far as the appeal concerns an order under s. 75 of the Waste Act ordering the cleaning of a littered area, appeals that may be filed do not, according to s. 139 of the Waste Act, have the ordinary suspensive effect on the implementation of the order. This means that a clean-up order can be implemented immediately, unless the appellate body stays the implementation. In addition to the party immediately concerned by the administrative decision, decisions issued under the Waste Act can also be appealed by a registered association or foundation mentioned in s. 138(1), para. 2, of the Waste Act.

As in the OPR Act, it is also according to the Waste Act possible for the public authorities to put administrative force behind their decisions.[84] The authorities, in this case mainly the regional Trade, Traffic and Environment Centre and the municipal environment committee, can, on the basis of s. 129 of the Waste Act, reinforce a prohibition or order based on a provision under the Waste Act with the threat of an administrative fine or a threat of the neglected measure otherwise being performed

---

[83] Kosola *et al.*, note 2 at 20.

[84] Concerning administrative force in the environmental area, see E. Hollo *Ympäristönsuojeluoikeus* (WSOY, Helsinki, 2001) 386–413.

at the defaulter's expense in the manner provided by the Administrative Fines Act 1990 (Finland). Administrative force is a suitable course of action in situations where the invitations and requests of the public authorities do not produce concrete results or where it is urgent to restore the situation so that it is in line with the law. In practice, public authorities have at their disposal few funds, which is a complication for using administrative force. The threat that a measure that has not been carried out is executed at the expense of the faulty party is, in the first instance, realised by work done at the expense of the municipality, the association of municipalities or the state, depending on whether the decision has been made by a municipal or state authority.[85] Because of the scarcity of public funds for this kind of work, the mechanism is not used very often. The expenses for the works done are, in the second instance, recovered from the responsible party in the manner established for the enforcement of orders for the recovery of taxes and fees. The use of administrative force is mainly possible in situations where the responsible party has funds of its own or where the vessel has some economic value.[86]

The removal of wrecks (and presumably also their cargoes) can be at least partly funded by the state, which means that the municipality is probably more easily persuaded to take action against wrecks that can be considered litter. Ultimately, of course, this means that the taxpayer of the municipality and of the state may become the final payer of the bill.

The removal of the wreck as waste management work is premised on the fact that the object is litter that has been left in the environment or that it is some other discarded object or substance that causes detriment or risk.[87] The term "environment" is, in this context, a broad term, including litter in water areas and on the bottom of water areas. Therefore, a wreck left in the sea may be such a discarded object referred to in the act. The condition for the funding of the project as waste management work is that an individual municipality would be caused unreasonable expenses for the work or measures that need to be taken for the prevention or elimination of the hazards, harm or other consequences referred to in the provision. In administrative practice, it has been considered that the expenses for a municipality would be unreasonable if they exceed 2.5 to 15 euros per inhabitant.[88]

---

[85] Kosola *et al.*, note 2 at 22.

[86] Kosola et al., note 2 at 21. For the general procedural norms concerning administrative force, see Kosola *et al.*, note 2 at 21–22, 23. According to Kosola *et al.*, note 2 at 25, wrecks become a problem in a situation when the vessel does not have any identifiable owner or possessor to which the administrative force could be directed and in whom the obligation to remove the vessel illegally abandoned could be directed. Also, such situations are problematic when the owner of the vessel is known but has been found insolvent, which means that use of administrative force is normally not possible.

[87] Because a wreck is likely to contain numerous hazardous substances, the removal and scrapping of a wreck is normally considered as an activity for which a so-called environment permit is required on the basis of s. 28 of the Environment Protection Act (86/2000). The consequence of this is that the removal and scrapping of the ship may be protracted.

[88] Kosola *et al.*, note 2 at 26–27. The level varies very much because of the very different economic capacity of the municipalities. Poor municipalities reach the upper limit sooner than rich municipalities.

Table 7.2 Removal of Wrecks by Municipalities

Munici-pality	Object	Year	Cost €	Funding share of the state	Funding share of the municipality	The number of inhabitants in the municipality	€/ inhabitant
Dragsfjärd	m/s *Kalana* removal & scrapping	1995	59,000	89%	11%	3613	16
Kustavi	m/s *Dagny* rem.&scrapping	1997	13,000	63%	38%	1020	13
Uusikaup.	m/s *Kehrwieder* rem.&scrapp.	1997	76,000	67%	33%	17,479	4
Pargas	m/s *Aliot* removal & scrapping	1999	34,000	67%	33%	12,005	3
Kemi	*Martinkari barge*	1997	3000	50%	50%	24,122	0

The participation of the state in the expenses takes place under the condition that the state budget contains funds for the purpose, and the same should be true for the municipality: the local authorities cannot act without funds that have been allocated for the purpose. This, of course, means that waste management works are normally not instantaneous, but are instead performed in the mid-term to long-term perspective, after the necessary funds have been budgeted for in the budget of the relevant public authority. For the state, the regional Trade, Traffic and Environment Centre decides on the participation of the state in the project and the works upon the initiative of the municipality. The state can decide to do the works itself or have the works done by some external contractor or participate in other ways in the expenses of the project. Depending on the way in which the waste management works are carried out, the regional Trade, Traffic and Environment Centre makes an agreement about the work either with the municipality or with the party which has the duty to clean up.[89] Apparently, the agreement referred to in the opening paragraph of this article concerning the removal of the ketch *Margona* was a deal of this sort. The table below gives an overview of practical instances in which wrecks have been removed by using funds from the state to supplement municipal expenses Table 7.2. It also indicates that the problems created by wrecks are, generally speaking, not very big – although they may be considerable for an individual municipality[90]:

---

[89] According to Kosola *et al.*, note 2 at 28, the state and the responsible party with the duty to clean up conclude the agreement on waste management works if the possessor of the area is liable for the expenses and the participation of the municipality is not necessary. In contrast to an agreement between the state and the municipality, which is an administrative agreement, an agreement between the state and the responsible private party might be a contract under private law, because the Waste Act and the Waste Decree do not create any explicit legal basis in the realm of public law for an agreement between the state and the private party.

[90] Source: Suomen ympäristökeskus 2004, as quoted in Kosola *et al.*, note 2 at 26.

There is no fixed share of the total cost for the state, but it seems that a fifty-fifty share is a starting point for the sharing of the cost with the municipality and there is the possibility to increase the share of the state.[91]

7.6 Concluding Remarks

There are different forms of legal obligations on public authorities that force them to take action against wrecks and cargo, including an obligation to remove if need be. This obligation exists on the basis of law for reasons connected to the environment and pollution. The obligation is sustained to some extent by s. 20 of the Constitution, but exists also for reasons of traffic safety and safety at sea. Therefore, the image created by the media that the legal basis for actions against wrecks is insufficient is incorrect. However, the insufficiency of public funds may restrict the removal of wrecks.

Wrecks may cause effects that are unlawful with reference to one or several Acts. The public authorities overseeing the implementation of each law have a duty to interfere in the factual situation in order to correct it. The public authorities have several alternatives that they can use. In the beginning, the public authorities may use soft means, that is, to request the person or company in charge of the vessel removes it or they make a stronger request to that effect. The next level is constituted by the different methods of exercise of administrative force by which a public authority can, for instance, create a duty for the owner to remove the vessel or wreck. Pressing criminal charges against the culprit is normally an ultimate form of reaction. Among these forms of reaction, the public authority shall choose the one which is the best fit for the situation at hand and the measures must be dimensioned with a view to the seriousness of the situation and in accordance with the given reasons and the expected consequences. It is therefore not possible to give categorical and schematic rules on how a particular situation should be dealt with.[92]

The public authorities that are obliged to act are the Finnish Environment Institute, the Traffic Authority, and the regional Trade, Traffic and Environment Centre or the municipal environment committee. They are different depending on the piece of law regulating the matter and their actions are differently fashioned in national law. However, international law is prominently featured as a basis for national law.

It appears as if the Finnish Environment Institute had a primary duty under the law to act in situations envisioned by the OPR Act, enacted against the background

---

[91] Kosola *et al.*, note 2 at 27–28. Based on s. 35 of the Waste Act and the above table, it is possible to disagree with Wetterstein's note 68 at 333, when he maintains that ultimately, the state will bear the burden of expenses. Instead, the starting point is that the municipality covers the expenses, that the state can pick up a good portion of the cost and that ultimately, the responsible party is subjected to attempts to recover the costs.

[92] Kosola *et al.*, note 2 at 20.

of the 1969 Intervention Convention. On the basis of the OPR Act and after an incident, the Finnish Environment Institute must at least consider taking measures against wrecks and cargo and should give reasons for not taking measures. Although it has not done so yet, the Institute *may* order rescue and other measures, but it does not have to do so. This means that the range of its discretion is very broad, although the original connection through the Intervention Convention to the measures taken in relation to *Torrey Canyon* is not visible in the 2009 Government Bill to the OPR Act in the manner outlined in the 1978 Government Bill to the former PPS Act. However, the powers of the Institute are not, after all, altogether dictatorial and arbitrary because of the different considerations that arise on the basis of law surrounding the decision-making situation. However, if the Institute has, after an incident, assessed the risk and not reacted and the environment has been harmed due to the incident, the Institute will have failed to fulfil the obligation that the OPR Act establishes, thus making it possible to attach negative legal consequences to a public authority on the basis of their inaction.[93]

Concerning safe traffic in fairways, the Maritime Act creates duties for the ship-owner that are, to a great extent, based on the 2007 Nairobi Convention, which brings clarity to the situation by containing explicit rules on governmental measures in relation to wrecks. Hence in principle, the maritime authorities should be able to take action against hazardous wrecks that are found in traffic routes, which means that the Traffic Authority appears to have a secondary responsibility after the ship-owner to remove wrecks that may become a primary one depending on the exigencies of the case. The explicit intention expressed in the Maritime Act is that the provisions concerning wrecks shall not be applicable in situations where the OPR Act is applicable, so a certain order of precedence is created for governmental reactions.

Finally, the Waste Act seems to be the piece of law that has been used the most in taking care of wrecks. On the basis of the Waste Act, the municipality in which the wreck is situated is ultimately responsible for measures in a tertiary instance (where the shipowner, the operator of the ship or the captain of the ship have antecedent obligations), but the state can participate in funding a removal and scrapping project. In spite of this positioning of the obligation, the public authorities maintain the right to seek the recovery of the costs from the responsible party. In relation to the OPR Act and the Maritime Act, it seems that the Waste Act is clearly a piece of "last instance legislation" and would be activated a relatively long time after the creation of a (non-hazardous) wreck or the sinking of cargo, when the operation of the other pieces of law may already have ceased. However, the Waste Act of mainland Finland does not apply in the autonomous Åland Islands, where the law of the autonomous entity governs wrecks and cargo that have become waste.

During the past decade, the Finnish law-maker has introduced improvements into the law concerning wrecks and cargo. However, national law in Finland does

---

[93] At the writing of the article, there are plans to transfer the tasks of the Finnish Environment Institute with regard to the oil pollution response to the Finnish Coast Guard. The transfer may take place during 2019.

not spell out the most radical forms of government action that might have to be considered in very serious situations, comparable to, for instance, the *Torrey Canyon* disaster, where the air force was used to bomb the wreck. Also, in regulating wrecks, current national legislation leaves open, at least to some extent, what should happen with, for instance, cargo that sinks in a fairway, but the broad understanding in the Nairobi Convention of what a wreck is should also be sufficient to cover cargo. In addition, while the law is increasingly making provisions about the responsibility of the owner of the ship to cover the costs of removal of a wreck (for instance, through the requirement of insurance in the Nairobi Convention), situations nonetheless may emerge in practice where the costs are, at the last instance, falling on the public budget and thus the taxpayer, which is clearly not satisfactory, but perhaps not entirely avoidable.

Chapter 8
Subsea Gas Pipelines in the Baltic Sea Area – Civil Liability Issues

Peter Wetterstein

Abstract The chapter focuses on legal issues linked to subsea gas pipelines, which have received transnational relevance in the Baltic Sea region through the Nord Stream pipeline between Russia and Germany. It includes a brief review of the legal basis of offshore activities, more generally and in the Baltic Sea. The main focus of the text is on civil liability issues linked to damage involving such pipelines, covering anything from different damage scenarios to questions related to liable persons and compensable loss, but also includes some choice of law aspects to the questions. It is concluded that a number of regulatory gaps exist in this area, but that an international regulatory regime on civil liability matters linked to subsea pipelines might still not be necessary.

Keywords Subsea pipelines · Civil liability · Baltic Sea · Nord Stream · Environmental damage · Remedying obligations · Choice of law

8.1 Introduction

Subsea gas pipelines are increasingly used in connection with gas exploration, exploitation and production activities offshore.[1] Gas is transported through pipelines below the water surface to other installations, onshore processing or

---

[1] "Offshore" is usually taken to mean areas covering the territorial seas, the Exclusive Economic Zone (EEZ) and the continental shelf. Cf. art. 2(2) of Directive 2013/30/EU of the European Parliament and of the Council on safety of offshore oil and gas operations and amending Directive 2004/35/EC. In art. 2(3) "offshore oil and gas operations" are described as "all activities associated with an installation or connected infrastructure, including design, planning, construction, operation and decommissioning thereof, relating to exploration and production of oil or gas, but excluding conveyance of oil and gas from one coast to another". Offshore oil and gas operations comprise all activities related to exploring for, producing or processing oil and gas offshore. In addition to the extraction of oil and gas from the underground strata of the seabed, these activities include the transport of oil and gas by vessels and through offshore structures (pipelines or other installations

P. Wetterstein (✉)
Åbo Akademi University, Turku, Finland
e-mail: peter.wetterstein@abo.fi

© Springer International Publishing AG, part of Springer Nature 2018
H. Ringbom (ed.), *Regulatory Gaps in Baltic Sea Governance*, MARE
Publication Series 18, https://doi.org/10.1007/978-3-319-75070-5_8

147

storage facilities. Furthermore, subsea pipelines are gaining growing importance as means of the transport of gas between different countries (cf. the Nord Stream project, *infra*).[2]

These activities also mean the growing risk of gas leaks, explosions and other accidents. For instance, trawling and dredging operations, vessels dropping/dragging anchors onto pipelines or submarines (or other submersible crafts) colliding with them, which may cause severe damage. Furthermore, errors or omissions by manufacturers, assemblers, repairers etc. of the pipelines or other technical or equipment failures may result in damage. The variety of harm and damage caused by such incidents may be great. In addition to "traditional" loss or damage, like personal injury, property damage and economic losses, gas releases may harm the marine environment and cause significant damage to eco-systems in the water and to protected marine species and natural habitats.[3] Also shipping, fishing and tourism sectors and coastal economies may suffer losses.

Consequently, questions of liability and compensation arise: Who has to pay for harm done and damage caused? What are the requisites for liability? Which losses and damages are compensable? How is the compensation to be calculated? Are there any insurance and financial securities? And so on. These issues are further complicated by the fact that most subsea gas pipelines run through the exclusive economic zones (EEZs) of states and across continental shelves, where the coastal state's jurisdictional competence is limited. Furthermore, an accident may concern more than one state. For instance, a gas release may affect many states, such as the states around the Baltic Sea, which are small and close to each other.[4]

This article will look into some of these issues from a civil liability perspective. The focus is on the Baltic Sea area[5] and especially the Nord Stream gas pipeline between Russia and Germany. However, before turning to civil liability rules and principles, including some choice of law issues, the article briefly reviews the legal basis of offshore activities.

---

above or below the water surface) to other installations, onshore processing or storage facilities.

[2] See generally A. Proelss, "Pipelines and protected sea areas", in R. Caddell and R. Thomas (eds), *Shipping, Law and the Marine Environment in the 21 st Century. Emerging challenges for the Law of the Sea – legal implications and liabilities* (Lawtext Publishing Limited, 2013), 276–292.

[3] See e.g. *DNV Recommended Practice DNV-RP-F107. Risk Assessment of Pipeline Protection* (October 2010), 19–34.

[4] Cf. P. Wetterstein, "Environmental liability in the offshore sector with special focus on conflict of laws (parts 1 and 2)" (2014) 20 *Journal of International Maritime Law* 33–34.

[5] The Baltic Sea is designated by the International Maritime Organization (IMO) as a "Particularly Sensitive Sea Area" (PSSA). On the legal status of PSSAs, see H. Honka, "Regulation of PSSA – Is it Effective?" in J. Schelin (ed) *Shipping & Environment – A Legal Perspective* (Poseidon Förlag AB, 2014), 23–33.

8.2 Public International Law Background

8.2.1 The Right to Lay Subsea Pipelines

The international legal basis of offshore activities, including the laying of subsea pipelines, is to be found in the UN Convention on the Law of the Sea (UNCLOS), adopted in 1982 and effective since 1994. UNCLOS provides a global framework for the rational exploitation and conservation of the resources in seas and oceans and the protection of the environment.[6] Both the European Union (1998) and Russia (1997) have ratified the Convention.

In the EEZ[7] and on the continental shelf[8] UNCLOS grants the coastal state sovereign rights over activities related to the exploration and exploitation of natural resources above and on the sea-bed and in the subsoil.[9] Further, art. 56.1(b) specifies

---

[6] For the development of international law leading to the adoption of UNCLOS, see e.g. P. Birnie, A. Boyle and C. Redgwell, *International Law and the Environment* (Oxford University Press, 2009), 379–84, and U. Beyerlin and T. Marauhn, *International Environmental Law* (2011), 115–19. UNCLOS, which, in large part, is a mere codification of the then existing rules of relevant customary and conventional international law, providing for a balance of power between coastal states and flag states. The emphasis of UNCLOS is strongly on the protection and preservation of the marine environment, rather than on compensating for environmental losses. On this see e.g. J.I. Charney "The Marine Environment and the 1982 United Nations Convention on the Law of the Sea" (1994) 28(4) *The International Lawyer* 879–901.

[7] The area adjacent to the seaward edge of the territorial sea is the EEZ. According to UNCLOS art. 57, the EEZ shall not extend beyond 200 nautical miles from the baselines from which the breadth of the territorial sea is measured. The EEZ is not part of the state's national territory; it is an area subject to the special legal regime established in Part V of UNCLOS, under which the rights and jurisdiction of the coastal state and the rights and freedoms of other states are governed by the relevant provisions of the Convention (art. 55). Thus, the coastal state can only exercise sovereign rights in compliance with this legal regime. For more, see e.g. S. Much "The emerging international regulation of carbon storage in sub-seabed geological formations" in R. Caddell and R. Thomas (eds), *Shipping, Law and the Marine Environment in the twenty-first Century. Emerging challenges for the Law of the Sea – legal implications and liabilities* (Lawtext Publishing Limited, 2013), 263–265, and Proelss, note 2 at 283–287.

[8] According to UNCLOS art. 76.1, the continental shelf of a coastal state comprises the sea-bed and subsoil of the submarine areas that extend beyond its territorial sea throughout the natural prolongation of its land territory to the outer edge of the continental margin, or to a distance of 200 nautical miles from the baselines from which the breadth of the territorial sea is measured where the outer edge of the continental margin does not extend up to that distance. The outer limits of the continental shelf on the seabed either shall not exceed 350 nautical miles from the baselines from which the breadth of the territorial sea is measured or shall not exceed 100 nautical miles from the 2500 metre isobath, which is a line connecting the depth of 2500 metres (art. 76.5, cf. 76.6). For more details on the delimitation of the continental shelf, see e.g. M.A. Allain "Canada's claim to the Arctic: a study in overlapping claims to the outer continental shelf" (2011) 42 *Journal of Maritime Law and Commerce* 13–19.

[9] See UNCLOS arts. 56, 60, 77, 80 and 81. While a coastal state's sovereign rights in the EEZ also embraces living natural resources (e.g. fish) of the waters superjacent to the seabed and economic activities, such as the production of energy from the water, currents and winds, the coastal state's

that in the EEZ the coastal state has jurisdiction as provided in the relevant provisions of UNCLOS with regard to, *inter alia*, the establishment and use of artificial islands, installations and structures. However, *other* states, whether coastal or landlocked, enjoy, subject to the relevant provisions of the Convention (especially art. 58.3[10]), the freedom "of the laying of submarine cables and pipelines" in the coastal state's EEZ (art. 58.1). Furthermore, all states are entitled to lay such cables and pipelines on the continental shelf, in accordance with the provisions of art. 79.[11]

8.2.2   Obligations to Protect the Marine Environment

UNCLOS contains many restrictions and limitations on the free use of the seas – especially with regard to protecting the marine environment. The rules on the protection and preservation of the marine environment are written into Part XII of the Convention.[12]

The general obligation of contracting states is written into art. 192: "States have the obligation to protect and preserve the maritime environment". More specifically, according to art. 194, states shall take *all measures* consistent with the Convention that are necessary to prevent, reduce and control pollution of the marine environment from any source (including pollution from vessels, installations and devices).[13] Obviously, this obligation also covers subsea pipeline activities in the EEZs and on the continental shelves.[14]

---

rights over the continental shelf relates mainly to non-living resources (oil, gas, minerals, etc.) of the seabed and subsoil (art. 77.4). For details, see e.g. K. Hakapää *Uusi kansainvälinen oikeus* (Talentum, 2010) 399–403, 410–416.

[10] Art. 58.3 reads: "In exercising their rights and performing their duties under this Convention in the exclusive economic zone, States shall have due regard to the rights and duties of the coastal State and shall comply with the laws and regulations adopted by the coastal State in accordance with the provisions of this Convention and other rules of international law in so far as they are not incompatible with this Part."

[11] See Proelss, note 2 at 283–292 on the scope of jurisdiction of the coastal state in respect of offshore pipelines. Also note UNCLOS art. 112.1: "All States are entitled to lay submarine cables and pipelines on the bed of the *high seas beyond the continental shelf*". (Emphasis added.)

[12] In addition to UNCLOS, an extensive legislative framework attempting to prevent or mitigate pollution at sea and to reduce the degradation of the marine environment has been developing since the 1970s. See M.N. Tsimplis "Marine pollution from shipping activities" (2008) 14 *Journal of International Maritime Law* 101.

[13] This duty of due diligence extends to the marine environment as a whole, including the high seas, and it covers all sources of marine pollution, including land-based and shipping/offshore activities. See Birnie, Boyle and Redgwell, note 6 at 387–389. In addition, states shall co-operate on a global basis and, as appropriate, on a regional basis, directly or through competent international organisations, in formulating and elaborating international rules, standards and recommending practices and procedures consistent with UNCLOS, for the protection and preservation of the marine environment, taking into account characteristic regional features (art. 197).

[14] On the coastal state's rights and obligations regarding the protection of the marine environment in its EEZ and continental shelf, see Proelss, note 2 at 278–292.

Furthermore, arts. 204 and 206 of UNCLOS contain provisions on the monitoring of the risks or effects of pollution and on the control and assessment of the potential effects of activities. Section 5 contains provisions on the obligations of states to adopt *international* rules and *national* legislation to prevent, reduce and control the pollution of the marine environment. Regarding pollution from seabed activities, reference is especially made to art. 208.[15]

However, neither UNCLOS nor the *other* international treaties on preventing, reducing and controlling marine pollution (e.g. the MARPOL Convention[16] 1973/78, the OPRC Convention[17] 1990, the OSPAR Convention[18] 1992 and the Helsinki Convention[19] 1992) provide for remedying responsibilities and the allocation of liability in relation to environmental damage (or other damage) caused by or in connection with subsea pipeline activities.[20] Therefore, our interest turns to international civil liability regimes and to national rules and principles on liability and compensation.

8.3 Civil Liability – Rules and Principles

8.3.1 General

There is no *international* regulatory framework in force specifically dealing with liability and compensation for damage or loss caused by gas exploitation or the use of gas pipelines or other installations. The lack of an international (or regional[21]) regime to regulate such liability could be seen as a deficiency in international law.[22]

---

[15] According to art. 208.1, which covers pollution from seabed activities subject to national jurisdiction, "[C]oastal States shall adopt laws and regulations to prevent, reduce and control pollution of the marine environment arising from or in connection with sea-bed activities subject to their jurisdiction and from artificial islands, installations and structures under their jurisdiction, pursuant to arts. 60 and 80". Such laws and regulations shall be no less effective than international rules, standards and recommended practices and procedures (art. 208.3).

[16] International Convention for the Prevention of Pollution from Ships. The entry into force of MARPOL has substantially reduced operational pollution from all types of vessels, see Birnie, Boyle and Redgwell, note 6 at 412 with references.

[17] International Convention on Oil Pollution Preparedness, Response and Co-operation.

[18] Paris Convention for the Protection of the Marine Environment of the North-East Atlantic.

[19] Convention on the Protection of the Marine Environment of the Baltic Sea Area.

[20] It should be noted that art. 235 of UNCLOS requires states cooperate on the development of international law relating to responsibility and liability for the compensation for damage caused by the pollution of the marine environment, as well as, where appropriate, the development of criteria and procedures for the payment of adequate compensation, such as compulsory insurance or compensation funds.

[21] "Regional" means that an instrument applies in the marine environment of more than two states.

[22] The legal situation is similar in the offshore oil exploration and exploitation sector. See Wetterstein, note 4 at 33–36.

As said, accidental blowouts and releases from offshore activities in the exclusive economic zones and on the continental shelves (and also the high seas) may have serious cross-border impacts.

Within the European Union we find legislation of some relevance on the liability issues mentioned; *for example,* Council Directive 85/374 on product liability (amended by Directive 1999/34/EC), EU Directive 2004/35 on environmental liability with regard to the prevention and remedying of environmental damage (ELD) and the Waste Framework Directive 2008/98/EC.[23] All states around the Baltic Sea, with the exception of Russia, are members of the EU.

However, as these EU instruments only cover liability issues to a limited extent, interest is directed towards national rules and solutions. Legal regimes and rules covering issues of liability and compensation related to gas pipelines (and generally offshore activities) vary significantly in different countries and regions. The variations embrace issues such as the basis and extent of liability, person (s) liable, damage and losses that are to be compensated, the extent of remedying obligations, financial security requirements, etc.[24]

Therefore, when considering liability issues related to subsea gas pipelines, the question of the applicable substantive law becomes topical. The outcome of an "international"[25] matter concerning liability and compensation is dependent on the legal norms applied by the court/authority seized with the case (and this is of importance for the choice of *forum* where the case is to be tried; see the Brussels I-Regulation[26] for the EU). As the focus of this article is on the Baltic Sea region, the Rome II-Regulation on the law applicable to non-contractual obligations[27] is of significance.

---

[23] This legislation replaces Directive 2006/12/EC of the European Parliament and of the Council of 5 April 2006 on waste.

[24] See e.g., K. De Smedt, M. Faure, J. Liu, N. Philipsen and H. Wang *Civil Liability and Financial Security for Offshore Oil and Gas Activities, Final Report* (2013), 82–154, 374–80. See also P. Wetterstein "Remedying of Environmental Damage Caused by Shipping" in (eds) J. Basedow, U. Magnus and R. Wolfrum *The Hamburg Lectures on Maritime Affairs 2009 & 2010* (2012) 179–192.

[25] Here is meant a "trans-boundary legal relationship" with connections to more than one legal system. See Wetterstein, note 4 at 97.

[26] Regulation (EU) No 1215/2012 of the European Parliament and of the Council of 12 December 2012 on jurisdiction and the recognition and enforcement of judgments in civil and commercial matters, which applies from 10 January 2015, with the exception of arts. 75 and 76, which apply from 10 January 2014 (art. 81).

[27] Regulation (EC) No 864/2007 of the European Parliament and of the Council of 11 July 2007 on the law applicable to non-contractual obligations. The Regulation was adopted on 11 July 2007 and entered into force on 20 August 2007. It entered into application on 11 January 2009 (with the exception of art. 29, which applied from 11 July 2008) and applies to events which occurred after the date of its entry into force. However, in *Homawoo v GMF Assurances SA* (C-412/10) 6 September 2011 the CJEU confirmed that the Rome II Regulation will apply to determine the governing law of non-contractual obligations only where the events giving rise to damage occurred after 11 January 2009.

8.3.2 Choice of Applicable Law

In cases of non-contractual obligations, like torts, many countries apply the principle of *lex loci delicti commissi*, that is, the legal norms pertaining in the country where the *event* giving rise to the damage occurred shall be applied regarding claims for damages. However, there have been varying interpretations/applications of this principle in cases of cross-border impacts.[28] Opinions have been expressed, for instance, in the Nordic countries, that the person suffering damage should be able to choose between the legal rules of the country of event and the country of damage (*lex loci damni*), according to whichever is more advantageous.[29]

Regarding liability issues related to subsea gas pipelines the *lex loci delicti commissi* rule would presumably lead to the application of the coastal state's law, provided that the event resulting in damage occurred in waters under that state's jurisdiction. But, as said, some courts may apply the law of the country where the damage occurred as an alternative.

This question has now been clarified for EU member states in the Rome II-Regulation, which applies "in situations involving a conflict of laws, to non-contractual obligations in civil and commercial matters" (art. 1.1). The Regulation is binding for member states[30] and it has *universal* application, that is, any law designated by it (*lex causae*) shall be applied whether or not it is the law of a member

---

[28] See e.g. Wetterstein, note 4 at 87 with references.

[29] See J. Cordes, L. Stenseng and P. Lenda *Hovedlinjer i internasjonal privatrett* (Cappelen Akademisk Forlag, 2010) 431, P. Wetterstein *Redarens miljöskadeansvar* (Åbo Akademis förlag, 2004) 463–464 with references, and more generally *Proposal for a Regulation of the European Parliament and the Council on the Law Applicable to Non-Contractual Obligations (Rome II)*, COM(2003) 427 final, where it is stated at 5: "Admittedly, the Member States virtually all give pride of place to the *lex loci delicti commissi*, whereby torts/delicts are governed by the law of the place where the act was committed. The application of this rule is problematic, however, in the case of what are known as 'complex' torts/delicts, where the harmful event and the place where the loss is sustained are spread over several countries. There are variations between national laws as regards the practical impact of the *lex loci delicti commissi* rule in the case of cross-border non-contractual obligations. While certain Member States still take the traditional solution of applying the law of the country where the event giving rise to the damage occurred, recent developments more commonly tend to support the law of the country where the damage is sustained. But to understand the law in force in a Member State, it is not enough to ascertain whether the harmful event or the damage sustained is the dominant factor. The basic rule needs to be combined with other criteria. A growing number of Member States allow a claimant to opt for the law that is most favourable to him…".

[30] EU member states must apply the Rome-II Regulation to proceedings commenced after 11 January 2009, and in those proceedings, its rules apply to events giving rise to damage occurring after 20 August 2007 (arts. 31–32).

state (art. 3). However, the respondent must be sued in a court of a member state (cf. the Brussels I-Regulation[31]).[32]

In addition to non-contractual obligations arising from land based activities, the Rome II-Regulation covers emissions and incidents in a state's sea territory and waters under its jurisdiction,[33] that is, basically its internal waters, the territorial sea, the EEZ and the continental shelf.[34] However, the Regulation determines only the *choice* of the applicable law, whereas the content and scope of application of that law is dependent upon the relevant state's jurisdictional competence to prescribe and enforce legislation which is restricted regarding the EEZ and the continental shelf (regarding subsea pipelines, reference is especially made to arts. 56.3 and 79.4 of UNCLOS[35]).

The *general rule* of the Rome II Regulation is *lex loci damni*:

> "Unless otherwise provided for in this Regulation, the law applicable to a non-contractual obligation arising out of a tort/delict shall be the law of the country in which the damage occurs irrespective of the country in which the event giving rise to the damage occurred and irrespective of the country or countries in which the indirect consequences of that event occur" (art. 4.1).[36]

Thus, the law applicable should be determined on the basis of where the *damage occurs*, irrespective of the country or countries in which the indirect consequences could occur.[37] Accordingly, in cases of personal injury or property damage, the

---

[31] See note 26. Given the complementarily between this Regulation and the Rome II Regulation, the universal nature of the latter is necessary for the proper functioning of the internal market as avoiding distortions of competition between EU litigants. See COM(2003) 427 final, 9–10.

[32] On the general objective of the Regulation, its scope of application and the notion of "non-contractual obligations", See Wetterstein, note 4 at 89–90 with references.

[33] Cf. art. 355 of the Treaty on the Functioning of the European Union (TFEU), which provides in essence that the Treaty and the instruments adopted thereunder apply to the member states and their European (and some overseas) territories.

[34] The Regulation lacks specific provisions as to its applicability on the continental shelf and the high seas. However, in Case-37/00 *Herbert Weber v. Universal Ogden Services Ltd.* the (former) ECJ treated installations on the continental shelf as part of the adjacent state for the purpose of jurisdiction over an employment dispute under the Brussels I Regulation. See also, Case C-347/10 *A Salemink v Raad van bestuur van het Uitvoeringsinstituut werknemersverzekeringen* (para. 35). See also B.O. Gram Mortensen and C.H. Schjøler "Sikkerhed og havvindmøller" (2013) 423 *MarIus* 71–80.

[35] On a coastal state's jurisdictional rights with respect to cables or pipelines on the seabed and subsoil, see Proelss, note 2 at 277–292.

[36] The scope of the applicable law is written into art. 15. The article covers issues including the basis and extent of liability, exceptions and limitations, damages and remedies, persons entitled to claim, and the extinguishment of obligations. On this see, A. Dickinson *The Rome II Regulation: The Law Applicable to Non-Contractual Obligations* (Oxford University Press, 2008) 568–591, A. Rushworth and A. Scott, "Rome II: choice of law for non-contractual obligations" (2008) *LMCLQ* 274, at 294, and R. Plender and M. Wilderspin *The European Private International Law of Obligations* (Thomson Reuters (Legal) Limited, 2009) 435–459.

[37] In COM(2003) 427 final, 11 it is stated: "The place or places where indirect damage, if any, was sustained are not relevant for determining the applicable law. In the event of a traffic accident, for example, the place of the direct damage is the place where the collision occurs, irrespective of

country in which the damage occurs should be the country where the injury was sustained or the property was damaged.[38] In most cases this means the country where the person suffering damage has his/her habitual residence.

There are two important *exceptions* to the main rule in art. 4.1. First, where the person claimed to be liable and the person sustaining damage both have their habitual residence[39] in the same country at the time when the damage occurs, the law of that country shall apply, that is, the *lex domicili communis* (art. 4.2).[40] Secondly, there is an "escape clause" in art. 4.3: "Where it is clear from all the circumstances of the case that the tort/delict is manifestly more closely connected with a country other than that indicated in paras. 1 or 2, the law of that other country shall apply".[41] This provision aims to bring a degree of flexibility, enabling a court to adapt the rigid main rule to an individual case so as to apply the law that reflects the "centre of gravity" of the situation.[42] However, since the clause generates a degree of unpredictability regarding the applicable law, it must remain exceptional.[43]

A significant *special* choice of law rule in the present context is the one included in art. 7: the law applicable to a non-contractual obligation arising out of *environmental damage*[44] or damage sustained by persons or property as a result of such damage shall be the *lex loci damni*, unless the person seeking compensation for

---

[38] financial or non-material damage sustained in another country". Regarding the indirect consequences of the event giving rise to damage, see A. Dickinson, note 36 at 313–18, and A. Rushworth and A. Scott, note 36 at 278–279.

[38] Recital (17). Consequently, where a single incident causes damage to occur in multiple states, art. 4.1 applies the law of each state to the damage that occurs there. On this issue see, e.g. Plender and Wilderspin, note 36 at 507–509. Furthermore, the present author agrees with the opinion of A. Dickinson, "Territory in the Rome I and Rome II Regulations" (2013) *LMCLQ* 86, according to which the connection between the relevant non-contractual obligation arising out of a tort/delict and the country whose law is to be applied under art. 4.1 is basically a factual one. He states that, "the member state court should seek to identify and locate the *facts* which represent the outward consequences of the defendant's conduct and then treat as the relevant 'damage' those consequences which are directly linked to that conduct and which are not indirect, in the sense that they simply reflect or follow from other consequences occurring in another country" (at 124). For more details on the difficulty in locating the relevant damage see 124–125.

[39] Regarding "habitual residence" see art. 23 of the Rome II Regulation, Dickinson, note 36 at 140–146, and Plender and Wilderspin, note 36 at 78–89.

[40] According to COM(2003) 427 final, 12: "[T]his is the solution adopted by virtually all the Member States, either by means of a special rule or by the rule concerning connecting factors applied in the courts. It reflects the legitimate expectations of the two parties."

[41] In the paragraph, it is further stated that, "[A] manifestly closer connection with another country might be based in particular on a pre-existing relationship between the parties, such as a contract, that is closely connected with the tort/delict in question".

[42] For more details see Plender and Wilderspin, note 36 at 536–541 and P. Stone *EU Private International Law* (Edward Elgar Publishing Limited, 2010) 386–389.

[43] COM(2003) 427 final, 12.

[44] According to recital (24) of the Regulation, "environmental damage" should be understood "as meaning adverse change in a natural resource, such as water, land or air, impairment of a function performed by that resource for the benefit of another natural resource or the public, or impairment of the variability among living organisms".

damage chooses to base his/her claim on the law of the country in which the event giving rise to the damage occurred.[45] This possibility to choose the applicable law resembles the earlier mentioned Nordic opinions regarding the interpretation of the *lex loci delicti commissi* principle.[46] However, a claimant who chooses the law of the event[47] cannot refer to the exceptions mentioned above (in art. 4.2–3).

Consequently, with regard to submarine cables and pipelines transporting oil or gas through coastal states' EEZs and continental shelves and causing damage (for instance, through explosions or other ruptures), arts. 4.1 or 7 apply accordingly. Also, issues of product liability may arise. Namely, art. 5 of the Rome-II Regulation contains a specific rule for non-contractual obligations in the event of damage caused by a *defective* product.[48] This rule is applicable to all defective products irrespective of the basis of liability or of the person sustaining the damage being a consumer or a business activity[49] and art. 5.1 contains a cascade system of connecting factors.[50] However, the Convention on the Law Applicable to Products Liability 1973 should also be noted in this context.[51]

---

[45] The motive for this rule is to be found in art. 191 of the TFEU which embraces, *inter alia*, the "polluter pays" principle. In COM(2003) 427 final it is stated, 19–20: "[The] exclusive connection to the place where the damage is sustained would also mean that a victim in a low-protection country would not enjoy the higher level of protection available in neighbouring countries. Considering the Union's more general objectives in environmental matters, the point is not only to respect the victim's legitimate interests but also to establish a legislative policy that contributes to raising the general level of environmental protection, especially as the author of the environmental damage, unlike other torts or delicts, generally derives an economic benefit from his harmful activity. Applying exclusively the law of the place where the damage is sustained could give an operator an incentive to establish his facilities at the border so as to discharge toxic substances into a river and enjoy the benefit of the neighbouring country's laxer rules. This solution would be contrary to the underlying philosophy of the European substantive law of the environment and the 'polluter pays' principle."

[46] See note 29.

[47] The question of when the person seeking compensation can make the choice of the law applicable should be determined in accordance with the law of the member state in which the court is seized (recital (25)).

[48] For the definition of a product and a defective product for the purposes of art. 5, see arts. 2 and 6 of Council Directive 85/374/EEC of 25.7.1985 on the approximation of the laws, regulations and administrative provisions of the Member States concerning liability for defective products, as amended by Directive 1999/34/EC of 10 May 1999.

[49] See C. Saf "Enhetliga regler för gränsöverskridande skadevållande händelser – Rom II-förordningen" (2009) *Europarättslig Tidskrift* 39.

[50] Namely, without prejudice to art. 4.2 (*lex domicili communis*), the law applicable to product liability shall be: "(a) the law of the country in which the person sustaining the damage had his or her habitual residence when the damage occurred, if the product was marketed in that country; or, failing that, (b) the law of the country in which the product was acquired, if the product was marketed in that country; or, failing that, (c) the law of the country in which the damage occurred, if the product was marketed in that country". A foreseeability clause (art. 5.1, para. 2) and an "escape clause" of the type included in art. 4.3 (art. 5.2) make the application of this conflict of law rule more complex. See Wetterstein, note 4 at 91–92 with references.

[51] The Rome II Regulation shall not prejudice the application of international conventions to which one or more member states were parties at the time when the Regulation was adopted and which lay down conflict of law rules relating to non-contractual obligations (art. 28.1). Such a convention

8.3.3 The Nord Stream Project as a Case Study

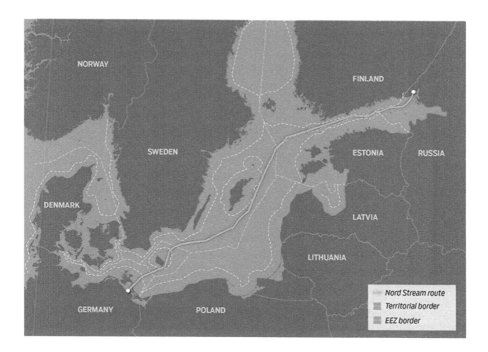

8.3.3.1 Background

The Nord Stream project consists of two parallel subsea pipelines transporting natural gas from Vyborg in the Russian Federation to Greifswald in Germany. The pipelines run along the bottom of the Finnish Gulf and the Baltic Sea and have a total length of 1222 kilometres, of which 375 kilometres is in the Finnish EEZ.[52] The pipelines, which are owned and operated by Nord Stream AG, were ready and in use 2011–2012.[53]

---

is the Convention on the Law Applicable to Products Liability, 1973, which takes precedence over the Regulation in the six member states which were parties to it when the Regulation was adopted on July 11, 2007. These states are Finland, France, Luxembourg, the Netherlands, Slovenia and Spain. Insofar as any issue not covered by the Convention is within the scope of application of the Rome II Regulation, the rules of the Regulation should be applied. See Plender and Wilderspin, note 36 at 563.

[52] In addition, the pipelines run through the territorial waters of the Russian Federation, Denmark and Germany, and through the EEZs of the Russian Federation, Sweden, Denmark and Germany.

[53] It should be noted that new parallel gas pipelines between the Russian Federation and Germany are being planned (Nord Stream 2). These pipelines should be ready and in use 2020. However, the project has met strong opposition. See https://en.wikipedia.org/wiki/Nord Stream. In addition,

The permission to carry out the project in the Finnish EEZ was given to the operator, Nord Stream AG, by the Finnish government on 5 November 2009. However, the decision did not discuss civil liability issues.[54] Thus, if there is an accident causing loss or damage in Finland (and in neighbouring countries[55]), interest is directed towards national rules on liability and compensation. According to the Act on Finland's Economic Zone (Laki Suomen talousvyöhykkeestä, 1058/2004), Finnish law is explicitly applicable to artificial islands, installations and structures in the EEZ (§ 17).[56]

8.3.3.2   Finnish Law on Liability and Compensation

General Rules

The Finnish Tort Act (Vahingonkorvauslaki, 412/1974) is the general law concerning non-contractual liability for damage or injury. It does not cover liability governed by *special* rules of law (Chap. 1 § 1). In the Tort Act liability is based on *fault*, including vicarious liability, such as an employer's liability for faults and the neglect of employees (Chap. 2 § 1 and Chap. 3), and covers personal injury, property damage and economic losses (also so-called *pure economic loss*[57] under certain conditions, Chap. 5 § 1[58]). The Act also contains a specific rule on the tort liability of

---

Baltic Connector Oy, a company owned by the Finnish state, is planning a gas pipeline with a length of 82 kilometres between Inkoo, Finland, and Paldiski, Estonia, to be ready in 2020. See https://en.wikipedia.org/wiki/Balticconnector

[54] The present author is unaware of any possible separate agreements/arrangements concerning liability and compensation. Cf. also UNCLOS art. 79.4 concerning the coastal state's right to establish conditions for cables or pipelines entering its territory or territorial sea, or its jurisdiction over cables and pipelines constructed or used in connection with the exploration of its continental shelf or the exploitation of its resources or the operations of artificial islands, installations and structures under its jurisdiction.

[55] Note the significance of the Rome II Regulation, *supra*.

[56] Cf. UNCLOS arts. 60.2 and 80.

[57] In the Finnish Tort Act there is no explicit definition of *pure economic loss*. However, both in legal literature and in court practice pure economic loss has been defined as "economic loss unconnected with personal injury or property damage" (cf. Chap. 5 § 1 of the Tort Act). A clear conceptual distinction is made in relation to *consequential loss*, that is, economic loss in connection with personal injury or property. This type of loss occurs, for instance, when an injured person loses income or a property owner loses profit because he is not able to use a damaged property. On the notion of *pure economic loss* in Finnish law, see e.g. Wetterstein, note 29 at 129–135, L. Sisula-Tulokas "Rena ekonomiska skador i skuggan av. sakskador" (2007) *Tidskrift utgiven av. Juridiska Föreningen i Finland* 415–427, and B. Sandvik "Ren förmögenhetsskada eller allmän förmögenhetsskada?" (2015) 8 *Svensk Juristtidning* 625–641 with references.

[58] The prerequisites for compensating pure economic loss are written into Chap. 5 § 1. Compensation is paid only if the loss is caused by 1) a criminal act, 2) by an administrative body in the exercise of its authority, or 3) if there otherwise exists a particularly weighty reason. On these prerequisites, see P. Wetterstein "Compensation for Pure Economic Loss in Finnish Tort Law" in P. Wahlgren (ed) *Tort Liability and Insurance* (Stockholm Institute for Scandinavian Law, 2001) 565–580, and P. Ståhlberg and J. Karhu *Suomen vahingonkorvausoikeus* (Talentum Helsinki, 2013) 321–330.

public authorities (the state, the municipalities, and other public organs) in the exercise of their authority. This liability rule is written into the Tort Act, Chap. 3 § 2, according to which a public organ is responsible for compensating damage caused during the exercise of its authority or power, but only to the extent that reasonable care has not been followed when taking into account the duties to be performed, the way in which the duties should be performed and the purpose of such duties. Fault or negligence on part of the person exercising public authority is, thus, a condition for liability.[59] As the notion *reasonable care* implies, wide room for discretion is left open to the courts.[60]

It is therefore evident that both operators of subsea gas pipelines (as well as other enterprises involved in such activities) and authorities in the performance of their supervision are open to liability under the Tort Act. In addition to the rules and provisions of the Act, general principles of tort law elaborated in court practice and legal literature are of relevance when evaluating the tort liability of pipeline operators. For instance, it has been held that operators of especially hazardous activities (the handling of explosives or other hazardous goods, the use of dangerous work machines and other technical devices, etc.) may be strictly liable for the injury or damage they cause.[61]

The Maritime Code (Merilaki, 1994/674) contains a provision on the vessel operator's ("laivanisäntä") vicarious liability in Chap. 7 § 1. The operator of a vessel is liable for loss or damage that the master, any member of the crew or a pilot causes through fault or neglect in the performance of their duties. He is also liable if the loss or damage is caused by any other person while performing work in the vessel's service on the order of the operator or the master, that is, the operator's liability also covers independent contractors involved in the operation of the vessel. Obviously then, this paragraph covers operators' liabilities when their vessels damage subsea pipelines (or other installations or structures).[62] Compensable damage is then regulated by the Tort Act and in Chap. 9 of the Maritime Code there are provisions on the limitation of liability.[63]

---

[59] It should be noted, however, that the state, the municipalities, and other public organs are liable, according to the Tort Act, Chap. 3 § 1, subpara. 2, for the negligent acts of their employees, civil servants, etc., when these are not acting in the exercise of public authority (vicarious liability).

[60] However, court practice does not seem to verify that there has been special restriction of liability in cases concerning the exercise of public authority. Consequently, one could argue that the provision, being insignificant, should be abolished. See L. Sisula-Tulokas, "Att ge och ta - ansvar för rena ekonomiska skador vid myndighetsutövning" (2010) *Tidskrift utgiven av Juridiska Föreningen i Finland* 313–320, Ståhlberg and Karhu, note 58 at 242, and P. Wetterstein "Det offentligas skadeståndsansvar – särskilt med hänsyn till sjöfartsförhållanden" (1999) 253 *MarIus*, at 5.

[61] See the Supreme Court's decisions KKO 1969 II 42, KKO 1990:55, KKO 1991:156, KKO 1994:122 and KKO 1995:108. Regarding legal literature, reference is made to, e.g. Ståhlberg and Karhu, note 58 at 170–176.

[62] See on the vessel operator's liability Wetterstein, note 29 at 33–55.

[63] See on limitation of liability Wetterstein, note 29 at 253–316.

In the Water Act (Vesilaki, 587/2011) there is a provision providing fault liability for owners of vessels causing economic loss ("edunmenetys") to third parties (Chap. 13 § 4). The scope of application of the Water Act extends to the Finnish EEZ (Chap. 1 § 4).

In the present context, the Product Liability Act (Tuotevastuulaki, 694/1990), which is based on the Directive 85/374/EC,[64] is also of interest. Defects in gas drilling, production and transport units, like the pipelines, may cause personal injury and property damage, thereby making producers[65] (and suppliers[66]) of defective products[67] strictly liable for the damage.[68] However, with regard to property damage, liability is limited to damage to property that "is of a type ordinarily intended for private use or consumption" and "was used by the injured person mainly for his own private use or consumption" (art. 9).[69] This restriction means that basically only private property is covered, for instance, damage to private vessels, jetties, beaches and water areas that are mainly used for non-commercial purposes.[70] However, the supplemental fault-based liability system under the Tort Act (*supra*) should be observed.

Environmental Liability Rules

Environmental Damage Act

The Environmental Damage Act, EDA (Ympäristövahinkolaki, 737/1994), is the general law concerning environmental impairment liability. The applicability of the EDA is determined by the description of the damaging activity and the types of compensable damage. According to § 1 of the EDA, compensation shall be payable for damage to the environment[71] caused by 1) the pollution of water, air, or land; or 2)

---

[64] Council Directive of 25 July 1985 on the approximation of the laws, regulations and administrative provisions of the Member States concerning liability for defective products. This directive has been amended by Directive 1999/34/EC of the European Parliament and of the Council of 10 May 1999.

[65] According to the Directive, "[P]roducer" means "the manufacturer of a finished product, the producer of any raw material or the manufacturer of a component part and any person who, by putting his name, trade mark or other distinguishing feature on the product presents himself as its producer" (art. 3.1).

[66] See art. 3.3 on suppliers' liability.

[67] For the purpose of the Directive, a "product" means "all movables even if incorporated into another movable or into an immovable" (in the present context especially note defective products incorporated into pipelines). 'Product' includes also electricity (art. 2).

[68] Exceptions to the liability are listed in art. 7.

[69] See also *Henning Veedfald v. Århus Amtskommune* (C-203/99).

[70] Loss of life and personal injury are generally covered by the Directive (art. 9(a)).

[71] However, the concept of "environment" is not defined. During the preparation of the EDA it was considered not possible to do so. See, P. Wetterstein "Environmental Damage in the Legal Systems of the Nordic Countries and Germany" in M. Bowman & A. Boyle (eds), *Environmental Damage in International and Comparative Law. Problems of Definition and Valuation* (Oxford University Press, 2002) 232.

noise, vibration, radiation, light, heating, or smell; or 3) other comparable distur-bance.[72] The EDA is applicable only to pollution from a *specific area* (harmful activity performed on land or water), and thus the EDA does not, as a rule, cover pollution from *moving* means of transport, such as vessels.[73]

In addition to compensation for personal injury and property damage,[74] the EDA covers pure economic loss,[75] provided that it is not insignificant (§ 5 para. 1).[76] More interestingly, § 6 of the EDA authorises the authorities[77] to claim *reasonable* (by reference to the disturbance or the risk of disturbance and the benefit of the restora-tion measures)[78] costs from the person(s) liable for measures undertaken to *restore*[79] the environment,[80] in addition to a private person whose individual rights have been infringed.[81] Under *public* and *administrative* laws the authorities often have a duty to take measures to protect and restore the environment.[82] The provision of the EDA helps to clarify the question of the responsibility to pay for these measures.

According to EDA § 7, strict liability is laid upon the *operator*, that is, the person who carries out the activity that causes the environmental damage. Furthermore, persons comparable with an operator (taking into consideration control, financial

---

[72] Regarding these requisites, see the extensive analysis by B. Sandvik *Miljöskadeansvar* (Åbo Akademis förlag, 2002) 123–174.

[73] But the maintenance of traffic areas, such as fairways and ports, is covered (§1 para. 2).

[74] Compensation for personal injury and property damage is payable according to Chap. 5 of the Tort Act. In this respect, the EDA does not introduce any changes.

[75] See note 57.

[76] Damage caused by a criminal act is always compensated. Furthermore, note EDA § 5 para. 2, which covers compensation for non-economic loss, for instance, inconvenience in exercising a right. However, compensation for non-economic loss, for instance, discomfort because of noise, smell, vibration, etc., is problematic because of the challenges of evaluating the damage. Therefore, it is provided in the EDA that such compensation shall be awarded to a *reasonable* amount. It is to be noted, however, that this kind of damage often has consequences of an economic nature, for instance, in the form of reduced market value for real estate, and the claimant is then, in certain circumstances, entitled to compensation for such economic loss.

[77] The term "authorities" covers both state and municipal authorities performing environmental protection.

[78] On the question of reasonableness, see Wetterstein, note 29 at 193–194 with references. Furthermore, the rules on "limits of tolerance" may also restrict the right to restoration costs. According to EDA § 4, damage to the environment is recoverable only if it is not reasonable to tolerate the disturbance taking into account, among other things, the local circumstances, the situ-ation as a whole that led to the disturbance, and how common this kind of disturbance is in com-parable circumstances. The obligation to tolerate disturbances is not applicable to personal injury or more significant property damage. Neither does it affect damage caused by criminal or inten-tional behaviour. On the "limits of tolerance", see Wetterstein, note 29 at 85–90.

[79] On restoration, see Wetterstein, note 29 at 234–241.

[80] See note 71.

[81] On the private person's right to claim restoration costs, see, e.g. Wetterstein, note 29 at 186–193.

[82] See Wetterstein, note 71 at 235, note 64. See also Ympäristövaliokunnan mietintö 3/2009 vp, Hallituksen esitys ympäristölle aiheutuvien vahinkojen korjaamista koskevaksi lainsäädännöksi, 2–3.

arrangements, etc.) also have strict liability, for instance, a parent company could be held liable for the activities of its subsidiary.[83] If the environmental damage is caused by two or more persons, they will be jointly and severally liable.[84] Consequently, the liability rules of the EDA could be of relevance to the operators (and comparable persons) of the Nord Stream project.

EU Directive 2004/35

Of relevance in the present context is EU Directive 2004/35 on environmental liability with regard to the prevention and remedying of environmental damage (ELD), which was adopted on 21 April 2004 and became fully binding on 30 April 2007. Thus all states around the Baltic Sea, with the exception of Russia, are bound by the ELD.[85] The Directive has been implemented in Finnish law mainly by the adoption of the Act on Remedying Certain Environmental Damage, (Laki eräiden ympäristölle aiheutuneiden vahinkojen korjaamisesta, 383/2009), the Water Act (mentioned above) and through amendments of the Nature Conservation Act (Luonnonsuojelulaki, 1096/1996) and the Environmental Protection Act (Ympäristönsuojelulaki, 86/2000).[86]

The ELD covers environmental damage and the imminent threat of such damage[87] caused by any of the occupational activities[88] listed in Annex III, which contains references to EU legislation.[89] Thus the ELD is of relevance also to activities related to subsea gas pipelines. If there is an emission or incident causing "environmental damage", the provisions of the ELD (as transposed into Finnish law) may be

---

[83] See the Finnish Government Bill 165/1992, 26–27. Regarding the liable person, see the Supreme Court's decisions KKO 1999:124, KKO 2001:61, KKO 2011:62 and KKO 2012:29. Also the transferee of an activity is liable if he knew or should have known about the damage or disturbance (or the risk of it) at the time of the transfer.

[84] However, a person whose apparent contribution to the damage is small cannot be held responsible for damage caused by others (§ 8). It will remain for the courts to decide what counts as "small".

[85] The EU member states were given time until that date to bring into force the legislation necessary to comply with the Directive (art. 19.1). Its implementation by all member states was completed by June 2010. On the legislative history of the ELD, see e.g. K. De Smedt "The Environmental Liability Directive: the directive that nobody wanted – Part I" (2015) 23 Environmental Liability 167–170. The ELD is currently being reviewed under the European Commission's REFIT programme (Regulatory Fitness and Performance programme).

[86] Laws 384/2009 and 385/2009.

[87] According to art. 2.9, "imminent threat of damage" means "a sufficient likelihood that environmental damage will occur in the near future".

[88] In the Directive "occupational activity" means "any activity carried out in the course of an economic activity, a business or an undertaking, irrespectively of its private or public, profit or nonprofit character" (art. 2.7).

[89] These activities include, inter alia, waste management operations, manufacture, use, storage, processing, filling, release into the environment and onsite transport of dangerous substances as defined in art. 2(2) of Council Directive 67/548/EEC (repealed by Regulation 2008/1272 EC), and the discharge or injection of pollutants into waters under the Directive 2000/60/EC.

applicable. Directive 2013/30/EU[90] expands the applicability of the ELD to cover the marine waters of member states as defined under the Marine Framework Directive 2008/56 EC.[91] Thus all installations in the waters of member states (including their EEZs and continental shelves) are covered.[92]

The notion of *environmental damage* covers a) *damage to protected species and natural habitats* (biodiversity),[93] which is any damage that has significant adverse effects on reaching or maintaining the favourable conservation status[94] of such habitats or species,[95] b) *water damage*, which is any damage that significantly adversely affects the ecological, chemical and/or quantitative status and/or ecological potential, as defined in Directive 2000/60/EC, of the waters concerned,[96] and c) *land damage*, which is any land contamination that creates a significant risk of human health being adversely affected as a result of the direct or indirect introduction, in, on or under land, of substances, preparations, organisms or micro organisms.[97] In the Directive *damage* has been defined as "a measurable adverse change in a natural resource or measurable impairment of a natural resource service which may occur directly or indirectly" (art. 2.2).[98]

The *operator*[99] of the activities listed in Annex III shall bear the costs for the preventive and remedial actions taken pursuant to the ELD (art. 8.1, strict liability

---

[90] Directive 2013/30/EU of the European Parliament and of the Council on safety of offshore oil and gas operations and amending Directive 2004/35/EC.

[91] See art. 38 of Directive 2013/30/EU.

[92] See the definition of "marine waters" in art. 3.1 of Directive 2008/56/EC.

[93] "Protected species and natural habitats" is explained in art. 2.3. Reference is made to the Wild Birds Directive 2009/147/EC (codified version of Directive 79/409/EEC) and the Habitats Directive 92/43/EEC.

[94] For the concept of "conservation status", see art. 2.4.

[95] The significance of such adverse effects is to be assessed with reference to the baseline condition, considering the criteria set out in Annex I to the ELD. "'Baseline condition' means the condition at the time of the damage of the natural resources and services that would have existed had the environmental damage not occurred, estimated on the basis of the best information available" (art. 2.14).

[96] With the exception of adverse effects covered by art. 4.7 of Directive 2000/60/EC.

[97] See art. 2.1.

[98] "Natural resource" means protected species and natural habitats, water, and land (art. 2.12) and according to art. 2.13, "'services' and 'natural resource services' mean the functions performed by a natural resource for the benefit of another natural resource or the public". On the conceptual issue "impairment of a natural resource service", see P. Wetterstein "A Proprietary or Possessory Interest: A *Conditio Sine Qua Non* for Claiming Damages for Environmental Impairment?" in P. Wetterstein (ed), *Harm to the Environment: The Right to Compensation and the Assessment of Damages* (Oxford University Press: 1997) at 48–50.

[99] In art. 2.6 "operator" is defined as "any natural or legal, private or public person who operates or controls the occupational activity or, where *this is provided for in national legislation* (my italics), to whom decisive economic power over the technical functioning of such an activity has been delegated, including the holder of a permit or authorisation for such an activity or the person registering or notifying such an activity". On the writing "operates or controls", see e.g. Statens offentliga utredningar (SOU) 2006:39. *Ett utvidgat miljöansvar. Delbetänkande av. Miljöansvarsutredningen* (2006) 103–108. Regarding specifically *offshore liability*, it should be

with some exceptions[100]). Occupational activities *other* than those mentioned in Annex III are subject to a fault-based regime (art. 3.1(b)). But such liability covers only damage and an imminent threat of damage to "protected species and natural habitats" (not water and land damage). However, a number of EU states, for example, Denmark, Finland, Latvia, Lithuania and Sweden, have included further activities not mentioned in Annex III under the scope of strict liability.[101] Regarding Finland, I restrict myself here to references only.[102]

The ELD has accepted the principle of liability for damage to the environment *per se*, that is, the "unowned" environment[103] (natural habitats, species of flora and fauna, water, land and soil etc.; cf. the notion of "environmental damage", *supra*). The environmental liability under the ELD is *exclusively* a liability vis-à-vis the public, that is, it aims to protect public rights.[104] It makes it possible for competent authorities to require that the preventive actions and remedial measures[105] are taken

---

noted that according to Directive 2013/30/EU, the licensee is liable for the prevention and remediation of environmental damage as defined in the ELD, caused by offshore oil and gas operations carried out by, or on behalf of, the licensee or the operator (art. 7). "Licensee" is defined as "the holder or joint holders of a licence" (art. 2(11)) and the "operator" as "the entity appointed by the licensee or licensing authority to conduct offshore oil and gas operations, including planning and executing a well operation or managing and controlling the functions of a production installation" (art. 2(5)). Recital 11 further clarifies that holders of authorisations for offshore oil and gas operations pursuant to Directive 94/22/EC are also the liable "operators" within the meaning of the ELD and should not delegate their responsibilities in this regard to the third parties contracted by them.

[100] The Directive does not cover environmental damage or an imminent threat of such damage caused by an act of armed conflict, hostilities, civil war or insurrection, or caused by a natural phenomenon of exceptional, inevitable and irresistible character. Furthermore, the operator is free of liability when he can prove that the environmental damage or imminent threat of such damage "(a) was caused by a third party and occurred despite the fact that appropriate safety measures were in place; or (b) resulted from compliance with a compulsory order or instruction emanating from a public authority other than an order or instruction consequent upon an emission or incident caused by the operator's own activities" (arts. 4.1 and 8.3). There are also defences introduced via transposition, e.g. the permit defence and the "state-of-the-art" defence (art. 8.4). Finland, among other member states, does allow the "permit" defence but not the "state-of-the-art" defence. On the varying decisions of the member states, see De Smedt, note 85 at 175.

[101] See Report from the Commission to the Council, the European Parliament, the European Economic and Social Committee and the Committee of the Regions under art. 14(2) of Directive 2004/35/EC on environmental liability with regard to the prevention and remedying of environmental damage, COM(2010) 581 final, 4.

[102] See the Act on Remedying Certain Environmental Damage § 1, the Nature Conservation Act § 5a, the Environmental Protection Act § 137 para. 2, § 176, and the Water Act Chap. 14 § 6.

[103] On this, see Wetterstein note 98, 30–32, 46–54.

[104] Traditional liability rules are normally concerned with proprietary or other *private* (individual) rights, as opposed to *public* (collective) rights, e.g. fishing rights in the sea and the right to use recreational areas. On these rights, see Wetterstein note 98, 30–43.

[105] According to recital (24): "Competent authorities should be in charge of specific tasks entailing appropriate administrative discretion, namely the duty to assess the significance of the damage and to determine which remedial measures should be taken". See also art. 11.2. On the role and obligations of the authorities, see e.g. M. Nesterowicz "The application of the Environmental Liability Directive to damage caused by pollution from ships" (2007) *LMCLQ* 113, 115–17.

the by the operator[106] and, if needed, to take these measures themselves,[107] and then recover all costs from the operator.[108] The ELD *does not* apply to cases of personal injury, damage to private property or to any economic loss and does not affect any right regarding these types of damage. Thus, unlike civil liability, it does not grant private victims any right of compensation. This is a significant limitation of the Directive's scope.

The ELD aims at fully restoring/compensating damage caused to natural resources and/or services. It emphasises the need for *in natura* restoration. When primary remediation does not result in fully restoring the environment, complementary remediation will be undertaken. The purpose of the latter remediation is to "provide a *similar level of natural resources and/or services, including, as appropriate, at an alternative site* (my italics), as would have been provided if the damaged site had been returned to its baseline condition".[109] In addition to these explicit provisions on alternative restoration, compensatory remediation shall be undertaken to compensate for the *interim loss* of natural resources and services pending recovery.[110] This compensation "consists of additional improvements to protected natural habitats and species or water at either the damaged site or at an alternative site". However, it does not provide financial compensation to members of the public.[111]

---

[106] For the liability mechanism to be effective, there needs to be one or more identifiable polluters, the damage should be concrete and quantifiable, and a causal link should be established between the damage and the identified polluter(s). See recital (13). However, causation issues (and the allocation of costs between multiple tortfeasors) was intentionally omitted from the scope of the ELD. See De Smedt, note 85 at 173, 175.

[107] The subsidiarity responsibility of competent authorities in the member states to take remedial measures when operators fail to do so has been left to the member states.

[108] See arts. 5–6.

[109] If possible and appropriate, the alternative site should be geographically linked to the damage site, taking into account the interests of the affected population (Annex II, 1.1.2). Complementary remediation can be used when the environment is so badly damaged that it cannot be restored in the particular location, or if complete restoration would take a very long period of time. As an example, if the damaged environment provides an essential ecological service, such as serving as a breeding ground or a habitat for a species requiring protection or a resting place for migratory birds or animals, then the environmentally useful remedy would be to provide an equivalent environment nearby. This could involve the acquisition and modification of a specific area of land or sea. See L. de La Fayette "The Concept of Environmental Damage in International Liability Regimes" in *Environmental Damage in International and Comparative Law – Problems of Definition and Valuation*, (eds) M. Bowman and A. Boyle (Oxford University Press: 2002) at 187.

[110] According to Annex II, 1(d), "interim losses" means "losses which result from the fact that the damaged natural resources and/or services are not able to perform their ecological functions or provide services to other natural resources or to the public until the primary or complementary measures have taken effect". H. Aiking, E. H P Brans and E. Ozdemiroglu "Industrial risk and natural resources: The EU Environmental Liability Directive as a watershed?" (2010) 18 *Environmental Liability*, 5–12, at 7, mention, as an example, that if a spill of chemicals results in significant damage to a number of acres of wetland and natural recovery is the most appropriate option, then during the recovery period some wetland services will be lost or impaired.

[111] Annex II, 1.1.3.

Regarding the complex issues of the identification of complementary and compensatory remedial measures and the choice of the remedial options when applying the ELD, I will cover the topics through references only.[112]

8.3.3.3 Summary of Civil Liability Rules Applicable to the Finnish EEZ

As said, the Nord Stream gas pipelines run 375 kilometres along the bottom of the Finnish EEZ. If applicable to damage caused by gas leaks, explosions or other accidents (cf. the Rome II Regulation), the relevant Finnish civil liability rules can be summarised as follows:

Finnish tort law is generally based on fault, including vicarious liability, both for employers and more extensively for the operators of vessels. In addition to fault liability, there are Supreme Court decisions holding the operators of especially hazardous activities strictly liable.[113] Compensation under the tort rules is awarded for personal injury, property damage, economic losses and, with restrictions, for pure economic loss. Thus, these general rules are of relevance to operators of subsea gas pipelines, vessel owners, supervisory authorities and others causing damage. Furthermore, producers and suppliers of defective products may be strictly liable under the rules on product liability.

With special regard to environmental impairment liability, the EDA is also applicable to subsea gas operations. The operators of such activities may be subject to claims for reasonable costs for the restoration of the environment (strict liability). In addition, the rules of the ELD, as transposed into Finnish law, may be applicable. As was said, these rules aim at fully restoring/compensating damage caused to natural resources and/or services through "primary", "complementary" and "compensatory" remediation. In contrast, the EDA contains no provisions accepting the idea that the *interim loss* of natural resources pending recovery should be compensated. Neither has such compensation been awarded in court practice. Furthermore, it is questionable whether restoration at an alternative site, so-called *alternative restoration*, may be approved and awarded under the EDA.[114] On the other hand (and in contrast to the ELD), the EDA also grants private victims right of compensation for personal injury, property damage and economic loss.

---

[112] See Wetterstein, note 4 at 44–45.

[113] There are also special rules of law providing for strict liability, e.g. regarding oil pollution from vessels, see P. Wetterstein "Environmental Liability of Operators in Ports - with Some Comments on the Concept of Environmental Damage" in H.Rak and P. Wetterstein (eds), *Environmental Liabilities in Ports and Coastal Areas - Focus on Public Authorities and Other Actors* (Institute of Maritime and Commercial Law, Åbo Akademi University: 2011) at 213–217.

[114] Sandvik, note 72 at 394, and Wetterstein, note 29 at 194 are in favor of such an application of the EDA.

8.4 Conclusion and Discussion

As noted, there is no international regulatory regime in place for specifically dealing with liability and compensation for damage or loss caused by gas exploitation or the use of subsea gas pipelines or other installations. In fact, the whole offshore sector lacks an international liability and compensation regime. Only *shipping* linked to the offshore sector is largely covered by international conventions, while other activities to a varying extent are governed by voluntary agreements (e.g. OPOL[115]), regional regimes (e.g. EU law), and national laws. However, the Legal Committee of the IMO has recognised that bilateral/regional agreements or arrangements are the most appropriate way to address these liability and compensation matters. According to the Committee, there is no compelling need to develop an international convention on this subject.[116]

Generally, and considering that offshore activities are more tied to regions and certain areas than shipping, I am inclined to agree with the views of the Legal Committee. The cross-border impacts of offshore activities usually strike certain geographical regions (for instance, the North and Mediterranean Seas and the Gulf of Mexico) that are in similar stages of economic, socio-cultural and legal development. Thus, it would seem to be easier to reach adequate solutions regarding liability and compensation, than when acting on the global arena (cf. shipping which operates globally). Such a regional arrangement, for instance, a separate European regime, ought to be comprehensive and cover both private claims for personal injury, property damage and economic losses, and public claims for damage to the environment (cf. the ELD mentioned above). One way to reach this objective could be to add an annex or protocol on civil liability to Directive 2013/30/EU on safety of offshore oil and gas operations. Ensuring prompt and adequate compensation would be essential, and this could be achieved through effective insurance/fund/risk pooling arrangements.[117]

---

[115] Offshore Pollution Liability Agreement 1974 (administered by the Offshore Pollution Liability Association Ltd., UK), effective in the North Sea area. This is a voluntary contractual regime entered into by a group of major oil companies involved in the exploitation and production of seabed mineral resources. See Wetterstein, note 4 at 34 with references.

[116] See Report of the Legal Committee on the Work of Its One Hundredth Session IMO LEG 100/14, 21–24.

[117] Reference is also made to the following statement in recital (63) to Directive 2013/30/EU: "Operators should ensure they have access to sufficient physical, human and financial resources to prevent major accidents and limit the consequences of such accidents. However, as no existing financial security instruments, including risk pooling arrangements, can accommodate all possible consequences of major accidents, *the Commission should undertake further analysis and studies of the appropriate measures to ensure an adequately robust liability regime for damages relating to offshore oil and gas operations, requirements on financial capacity including availability of appropriated financial security instruments or other arrangements. This may include an examination of the feasibility of a mutual compensation scheme...*" (my italics). De Smedt, Faure, Liu, Philipsen and Wang, note 24 at 173–226, 380–382, examine and discuss the interesting question of the extent to which risk pooling mechanisms could play an important role in compensating for damage that results from offshore incidents.

Turning now to subsea gas pipelines in the Baltic Sea, is there a need for more uniform liability rules related to such pipelines? A separate comprehensive EU regime for the whole offshore sector would, to a large extent, solve the issue (note, however, that Directive 2013/30/EU excludes "the conveyance of oil and gas from one coast to another"). While waiting for such a solution, what needs to/can be done?

As said, all states around the Baltic Sea, with the exception of Russia, are EU states. However, the EU legislation in force covers civil liability issues only to a limited extent. Directive 85/374 on product liability (with later amendments) is of relevance, and Directive 2004/35 on environmental liability, although mainly not being a civil liability regime, obliges operators of activities mentioned in the Directive to bear the costs for preventive and remedial actions. As regards the remedying of environmental damage, this is important, but the Directive does not grant private victims any right of compensation.

Cases of personal injury, damage to private property or economic loss are still governed by national rules on civil liability and compensation. Although there are significant divergences between the jurisdictions around the Baltic Sea, one can also find similarities. For instance, similar maritime laws (based partly on international conventions[118]) regulate liability and compensation for damage caused by vessels. Furthermore, there are groups of countries, like the Nordic states, with similar tort and environmental laws based on a long-standing legislative cooperation.[119] Concerning conflict of laws issues, the Rome II-Regulation has brought clarity and predictability to the EU.

Hence, I am not convinced that there is currently the need to develop uniform civil liability rules related to gas (and other) pipelines on the bed of the Baltic Sea. If there will be no comprehensive EU liability regime in place for the whole offshore sector in the future, and the need for rules concerning pipelines would arise, then some judicial actions could be considered, for instance, developing model laws on civil liability and compensation (cf. the ongoing efforts to harmonise the European law of obligations[120]) or reaching some kind of agreement/treaty on these issues between the states around the Baltic Sea (or between major gas companies/stakeholders[121]). Furthermore, questions of compulsory third party insurance or other financial securities could then arise.

---

[118] See e.g. Wetterstein, note 4 at 36–41 with references. Also Russia is a party to most of these conventions.

[119] These rules address rather well the civil liability issues discussed in the present article. Regarding environmental impairment liability, see Wetterstein, note 71 at 223–242. It should be noted, however, that the ELD has been thereafter transposed into the national laws of these countries.

[120] Principles, Definitions and Model Rules of European Private Law. Draft Common Frame of Reference (DCFR).

[121] Cf. the OPOL-agreement mentioned in note 115.

Chapter 9
Using the Continental Shelf for Climate Change Mitigation: A Baltic Sea Perspective

David Langlet

Abstract The chapter addresses legal issues relating to the use of the continental shelf for carbon storage in the Baltic Sea. Following an introduction to the technology, the focus is on inconsistencies within and between different regulatory layers in this novel yet fairly well-regulated field and on the ability of the legal system to manage the many interests and actors affected by such operations. The last part of the article assesses how well equipped international and EU law is to deal with conflicting interests and recent developments in this regard. While focusing on carbon storage, the significance of the analysis is more broadly applicable for any other marine activities that give rise to conflicting interests in the Baltic Sea region and beyond.

Keywords Carbon storage · Climate change · Continental shelf · Baltic Sea · Conflicting interests · Geoengineering · Environmental impact assessment · Marine spatial planning

9.1 Climate Change Mitigation and the Seabed

Climate change mitigation comes in many forms, with varying connection to the sea in general and the seabed in particular. A major strand of climate change mitigation is obviously activities aimed at reducing the anthropogenic emissions of greenhouse gases by transforming the energy system from a fossil-based one to one utilising renewable sources of energy. Among such sources of energy some may or must be deployed at sea. These include offshore wind farms as well as wave and tidal energy plants. All of these may have an impact on the seabed, e.g. through drilling and anchoring of structures.

Another way of mitigating climate change is to create or increase the capacity of carbon sinks, i.e. natural or artificial reservoirs that accumulate and store

D. Langlet (✉)
School of Business, Economics and Law, University of Gothenburg, Gothenburg, Sweden
e-mail: david.langlet@law.gu.se

© Springer International Publishing AG, part of Springer Nature 2018   169
H. Ringbom (ed.), *Regulatory Gaps in Baltic Sea Governance*, MARE
Publication Series 18, https://doi.org/10.1007/978-3-319-75070-5_9

carbon. Like forests, the biological life of the oceans may serve as a carbon sink. Their capacity to perform this 'good' may be enhanced by fertilizing the sea to stimulate the growth of plankton. Since this intentionally alters the natural environment by increasing the volume of biomass in the ocean, and thereby the amount of carbon dioxide sequestered by such biomass, ocean fertilization is seen as a form of geoengineering.[1] Ocean fertilization as a response to climate change is quite controversial.[2] It is also hardly an option in the Baltic Sea region due both to the fact that the sea is already more than sufficiently provided with most nutrients[3] and the fragile ecosystems of the area, which are already under severe pressure and often suffering from hypoxia or anoxia, i.e. lack or absence of oxygen.[4]

Other forms of geoengineering, such as solar radiation management, would not specifically affect the seabed.[5] The most significant impact on the seabed is likely to follow from the deployment of another sink-technique, namely carbon capture and storage (CCS), at least if the storage takes place offshore. CCS, which does not count as geoengineering as long as carbon dioxide is not captured from the atmosphere but rather in direct connection with the burning of fossil fuels,[6] is a collective name for a number of, partly alternative, techniques and methods which, when combined, allow carbon dioxide to be sequestered and thus prevented from reaching the atmosphere. Technically, it comprises three main stages: capturing the carbon dioxide (pre- or post-combustion), transporting it to a suitable storage site and final storage/disposal.[7]

From a legal point of view, CCS with offshore storage of carbon dioxide raises a host of issues, some of which it shares with sea-based renewable energy technologies such as wind farms. Some aspects of CCS, such as the use of pipelines for transporting carbon dioxide to suitable storage locations are also very similar to well established uses of the seabed that are not related to climate change mitigation, e.g. submarine oil and gas pipelines. This makes offshore CCS a good case for also studying regulatory problems pertaining to seabed utilisation more generally. But there are important regulatory issues which are distinct to CCS. To a large extent,

---

[1] On geoengineering generally, see The Royal Society *Geoengineering the climate: science, governance and uncertainty* (September 2009, Report 10/09).

[2] On the regulation of ocean fertilization, see Report of the Thirtieth Consultative Meeting of Contracting Parties to the London Convention and the Third Meeting of Contracting Parties to the London Protocol (9 December 2008) LC 30/WP.3. See also CBD COP 9 Decision IX/16: Biodiversity and climate change (9 October 2008) UNEP/CBD/COP/DEC/IX/16.

[3] In fact, nearly the entire Baltic Sea area is affected by eutrophication. M. Pyhälä et al. (eds), Eutrophication status of the Baltic Sea 2007–2011 - A concise thematic assessment, Baltic Sea Environment Proceedings No. 143 (HELCOM, Helsinki, 2014) 6.

[4] Extensive algal blooms can lead to severe oxygen depletion as a result of cellular respiration and bacterial degradation.

[5] On solar radiation management, see The Royal Society, note 1, Chap. 3.

[6] If carbon dioxide from the burning of biomass is sequestered, it could count as geoengineering since the carbon dioxide would then be part of the carbon dioxide cycle.

[7] For further details on the technology, see Intergovernmental Panel on Climate Change (IPCC) *IPCC Special Report on Carbon Dioxide Capture and Storage, prepared by Working Group III of the IPCC* (Cambridge University Press, Cambridge, 2005).

these relate to inconsistencies between regulatory levels or frameworks and to the effects of the uncoordinated developments of law in different fora.

In the following, legal complications relating to the use of the continental shelf for CCS operations will be discussed in some detail and focus on the effects of such inconsistencies as well as on the ability of the legal system, broadly speaking, to manage the many interests and actors affected by CCS operations involving the seabed. First, however, the technology as such requires a short introduction.

9.2   Carbon Capture and Storage – Technology in Context

As already noted, CCS comprises three more or less distinct steps: capture, transport, and storage. The capturing of carbon dioxide does not typically raise issues pertaining to the sea and its regulatory framework. Also, the origin of the carbon dioxide to be stored generally lacks relevance for the legal issues discussed here, although it may have implications for technical aspects such as pipeline design, and in relation to the legal and financial incentives that apply for using CCS in the first place. It should be mentioned, though, that in a Nordic perspective, where biomass plays an important role in most national energy systems, co-called bio-CCS, which involves storing carbon dioxide generated when combusting biomass, can be of particular interest.[8] Bio-CCS may be regarded as leading to "negative emissions" since it results, at least when seen over a period of some years or a few decades (depending on what kind of biomass is being used) in the removal of carbon dioxide from the atmosphere, rather than merely preventing the addition of carbon dioxide to it. Among others, the Swedish government has identified a potential for employing bio-CCS in the paper and pulp industry.[9]

As regards transport, there are two main options: ships and pipelines. These can be combined so that land-based pipelines take the carbon dioxide to seaside terminals where it is loaded onto ships. Ship transport is often seen as a preferable and perhaps even indispensable part of the ramp up phase of large CCS operations. Once the system is up and running, more industries will hopefully join and the volumes will rise to levels where pipeline transport is economically and practically the more feasible option.[10]

The geological storage of captured carbon dioxide can take place onshore as well as offshore in the seabed. The costs associated with sub-seabed storage are generally higher than for storage on land. However, offshore storage can still be the most

---

[8] On biogenic emissions in the Nordic region see S. Teir et al. *Potential for Carbon Capture and Storage (CCS) in the Nordic Region* (VTT Research Note 2556, 2010) available at <http://www.vtt.fi/inf/pdf/tiedotteet/2010/T2556.pdf>, at 23.

[9] Government Bill 2016/17:146 *Ett klimatpolitiskt ramverk för Sverige*, 34.

[10] N. Rydberg & D. Langlet *CCS in the Baltic Sea region – Bastor 2: Work Package 4 – Legal & Fiscal Aspects* Elforsk report 14:48 (2015) available at <http://www.elforsk.se/Rapporter/?rid=14_48_> at 30 with further references.

attractive or even only possible storage option. In some areas, including the Nordic region, the majority of the estimated storage capacity is found offshore.[11] Offshore storage may also be the most viable, or, indeed, the only viable option from a policy perspective mainly due to the issues of safety and, not least, public perception.[12]

The precondition for offshore storage in the Baltic Sea area varies significantly between regions and countries. For example, among the Nordic countries, Norway has very large offshore aquifer storage capacity, Denmark has a much smaller but still significant onshore and offshore capacity, whereas Finland lacks bedrock suitable for the geological storage of carbon dioxide.[13] Any carbon dioxide captured in Finland would thus have to be stored elsewhere. In the Baltic Sea good conditions for storing carbon dioxide have been identified in geological formations that stretch across the Latvian, Lithuanian and Swedish continental shelves, and to a lesser extent those of Poland and Russia (Kaliningrad).[14]

It should also be recognised that the mere existence of storage capacity within a state, including on its continental shelf, does not mean that it might not make more economic, and possibly technical, sense to use storage facilities elsewhere.[15] The need to attain a sufficient scale for any CCS project to make financial sense is also likely to necessitate the linking of emissions sources across national borders.[16]

This is reflected in EU infrastructure policy through the designation of the "development of carbon dioxide transport infrastructure between member states and with neighboring third countries in view of the deployment of [CCS]" as a "priority thematic area" under the guidelines for trans-European energy infrastructure.[17] Carbon dioxide transport infrastructure projects can thus qualify to become projects of European Common Interest.

In this context, it should be noted that whereas most states in the Baltic Sea area allow the geological storage of carbon dioxide, Finland and Estonia have opted to not allow such storage.[18] Sweden, on the other hand, has lifted a previous ban on storage in the seabed. Germany has restricted the annual quantity of carbon dioxide that may be stored to 4 Mt as a national total and 1.3 Mt per storage site.[19]

Undoubtedly, CCS is a controversial technology the use of which is both technically challenging and costly. Still, it has been described by the International Energy

---

[11] This applies e.g. in the Nordic region. See Teir et al., note 8 at 73.

[12] Intergovernmental Panel on Climate Change, note 7 at 34.

[13] Teir et al., note 8 at 73.

[14] *CCS in the Baltic Sea region – Bastor 2 Final Summary Report* Elforsk report 14:50 (2014) Chap. 2 available at <http://www.elforsk.se/Rapporter/?rid=14_50_>

[15] See e.g. *Carbon Capture and Storage in the Skagerrak/Kattegat region, Final report* (February 2012) available at <http://www.ccs-skagerrakkattegat.eu/news/tabid/60/Default.aspx>.

[16] See, e.g. the models/scenarios presented in ibid. and in Teir et al., note 8.

[17] Regulation (EU) No 347/2013 of the European Parliament and of the Council of 17 April 2013 on Guidelines for Trans-European Energy Infrastructure, OJ [2013] L115/39, Annex I.

[18] Report from the Commission on the implementation of Directive 2009/31/EC on the geological storage of carbon dioxide (25 February 2014) COM(2014) 99 final, 3.

[19] Ibid.

Agency (IEA) as a critical component in the portfolio of low-carbon energy technologies if governments are to undertake ambitious measures to combat climate change.[20] Its attraction also lies in the fact that CCS is the only technology available today that has the potential to mitigate climate change while preserving the value of fossil fuel reserves and the existing fossil fuel-related infrastructure.[21] Whether that is a good thing may of course be disputed, but it is evidently challenging to rid the whole economy of all significant sources of fossil carbon dioxide, not least the parts of industry where such emissions are linked to production processes rather than only to the generation of energy.[22]

Despite being identified as an essential part of the portfolio of climate change mitigation technologies,[23] CCS is facing several challenges, the most severe of which is probably the lack of a clear business case for investing in CCS in most jurisdictions.[24] In the EU this is largely a result of the EU emissions trading scheme for greenhouse gases (EU ETS) failing to push the price of carbon dioxide emissions from the so-called "covered sector" to levels that effectively influence industries' investment decisions.[25] But there are also legal and policy obstacles to CCS deployment, not least in the Baltic Sea region. In the following, only issues relating to the utilisation of the continental shelf from a Baltic Sea perspective or issues that otherwise directly impact on that utilisation will be discussed.[26]

---

[20] International Energy Agency (IEA) *Technology Roadmap: Carbon Capture and Storage* (IEA, Paris, 2013) available at <http://www.iea.org/publications/freepublications/publication/technology-roadmap-carbon-capture-and-storage-2013.html> at 5.

[21] Ibid., at 8.

[22] This applies with respect to steel, cement and parts of the chemicals industry. Teir et al., note 8 at 8.

[23] Referring to the EU's target of reducing greenhouse gas emissions by at least 80% by 2050, the EU Commission has found that "the 2050 target can only be achieved if the emissions from fossil fuel combustion are eliminated from the system, and here CCS may have an essential role to play, as a technology that is able to significantly reduce carbon dioxide emissions from the use of fossil fuels in both the power and industrial sectors". Communication by the Commission on the Future of Carbon Capture and Storage in Europe COM(2013) 180 final, 11.

[24] *The Global Status of CCS: 2013* (Global CCS Institute, Melbourne, 2013) 60; Communication by the Commission on the Future of Carbon Capture and Storage in Europe, note 23 at 16.

[25] The EU ETS is set up through Directive 2003/87/EC of the European Parliament and of the Council of 13 October 2003, establishing a scheme for greenhouse gas emission allowance trading within the Community and amending Council Directive 96/61/EC [2003] OJ L275/32. The Directive applies to emissions from the activities listed in Annex I and the greenhouse gases listed in Annex II. The activities covered include, inter alia, combustion installations, oil refineries, installations for the production of pig iron or steel, installations for the production of cement and industrial plants for the production of pulp from timber or other fibrous materials, paper and board. In most cases only activities with a certain output, expressed, e.g. as a minimum number of tonnes produced per day, are covered. All in all, more than 11,000 power stations and industrial plants are included. On the EU ETS, see D. Langlet and S. Mahmoudi *EU Environmental Law and Policy* (Oxford University Press 2016), section 11.2.

[26] On the regulatory aspects of CCS more broadly, see I. Havercroft, R. B. Macrory, and R. B. Stewart (eds) *Carbon Capture and Storage – Emerging Legal and Regulatory Issues* (Hart, Oxford and Portland, 2011).

9.3 CCS – General Regulatory Preconditions

9.3.1 Introduction

The Baltic Sea has nine coastal states: Denmark, Estonia, Finland, Germany, Latvia, Lithuania, Poland, Russia, and Sweden. Of these, eight are member states of the EU. Obviously, this makes the Baltic Sea area strongly influenced by EU law. However, when talking about the legal regime pertaining to the seabed it is inevitable to start the analysis with the United Nations Convention on the Law of the Sea (UNCLOS).[27]

Although being distinct systems of law, there is a strong functional link between UNCLOS and international law relevant to marine conditions more generally, and the legal system established by the EU. After a short discussion of the most pertinent features of UNCLOS, we will return to the effect of EU law and also the relationship between EU law and pertinent international law, including UNCLOS.[28]

This is not the place for a general discussion about the legal preconditions for the exercise of jurisdiction in different areas, i.e. maritime zones, and with respect to different activities at sea.[29] But before engaging in more detail with the legal preconditions for CCS operations in the Baltic Sea context, it is appropriate to briefly recall some features of the jurisdictional system at the core at the law of the sea and elaborate on their implications for CCS. The jurisdictional rules are determinative of who has the right to authorise or regulate CCS activities in different areas, as well as who is responsible for the diligent application of, e.g. international environmental law and, indirectly, EU law with respect to a specific activity.

9.3.1.1 Maritime Areas and Their Relevance to CCS

As is well known, the fundamental rules on activities at sea or otherwise directly relating to the seas are laid down in UNCLOS to which all the coastal states of the Baltic Sea as well as the EU are parties. UNCLOS defines and establishes the preconditions for the different maritime jurisdictional zones.

The starting point for any discussion about such zones is the baseline, which – as a default – is the low-water line along the coast, marked on large-scale charts officially recognised by the coastal state (normal baseline). However, where the coastline is deeply indented and cut into or, if there is a fringe of islands along the coast in its immediate vicinity, so-called straight baselines may be drawn by joining

---

[27] United Nations Convention on the Law of the Sea (Montego Bay, 10 December 1982) 1833 UNTS 3 (UNCLOS).

[28] See section EU LAW RELEVANT TO CCS: GENERAL PRECONDITIONS below.

[29] See instead e.g. D. R. Rothwell, A. G. Oude Elferink, K. N. Scott, T. Stephens (eds) *The Oxford Handbook of the Law of the Sea* (Oxford University Press, Oxford, 2015); and Y. Tanaka *The International Law of the Sea* (2nd edn, Cambridge University Press, Cambridge, 2015).

appropriate points on the coast or on certain elevations.[30] Straight baselines are widely used in the Baltic Sea.

The territorial sea starts on the seaward side of the baseline, which, where geography allows, can stretch out to 12 nm from the baseline. With the significant exception of the right to so-called innocent passage,[31] the sovereignty of the coastal state applies in the territorial sea more or less as it does on its land territory.[32] That means that in the territorial sea the coastal state has the exclusive right – although subject to obligations under international and EU law – to regulate any CCS activities. No drilling or injection of carbon dioxide into the seabed may take place here without the consent of the coastal state. The same goes for the construction of installations for such injection as well as for the laying and operation of pipelines for the transport of carbon dioxide.

Before engaging with the exclusive economic zone (EEZ), it should be noted that in the Baltic Sea there is no high seas, i.e. parts of the sea not included in the EEZ, the territorial sea or the internal waters of a state.[33] But it is still relevant to note that the so-called freedom of the high seas comprises not only a freedom of navigation and of overflight but also, inter alia, the freedom to lay submarine pipelines and cables.[34] This is relevant since, as we shall see, the freedoms of the high seas apply in principle, although subject to additional restrictions, also in the EEZ.

The EEZ is an area beyond and adjacent to the territorial sea. If an EEZ has been established, as is the case for all the coastal states of the Baltic Sea, it stretches from the outer limit of the territorial sea to a maximum of 200 nm from the baseline.[35] In the Baltic Sea the EEZ does not reach such a breadth since the distance between states with facing coasts does not allow for that.

In the EEZ the coastal state enjoys sovereign rights for the purpose of exploring and exploiting as well as conserving and managing living as well as non-living natural resources. This applies to the waters superjacent to the seabed as well as to the seabed and its subsoil. The coastal state also has sovereign rights with regard to other activities for the economic exploration and exploitation of the zone.[36] This includes the production of energy from water, currents and winds. Furthermore, the coastal state has jurisdiction in the EEZ, as provided for in the relevant provisions of UNCLOS, with regard to the protection and preservation of the marine environ-

---

[30] UNCLOS, arts. 5 and 7.

[31] The right to 'innocent passage' means essentially that ships from other states may navigate in this area as long as their passage is continuous and expeditious and they do not engage in activities prejudicial to the peace, good order or security of the coastal state. UNCLOS, arts. 17–19.

[32] UNCLOS, art. 2.

[33] UNCLOS, art. 86. To be exact, the archipelagic waters of an archipelagic state are also not part of the high seas.

[34] UNCLOS, art. 87(1). Additional freedoms are the freedom to construct artificial islands and other installations permitted under international law, subject to UNCLOS Part VI; the freedom of fishing, subject to the conditions laid down in UNCLOS Part VII, section 2; the freedom of scientific research subject to UNCLOS Parts VI and XIII.

[35] UNCLOS, arts. 55 and 57.

[36] Ibid., art. 56(1)(a).

ment, marine scientific research and the establishment and use of artificial islands, installations and structures.[37] Pipelines, which are not installations or structures in this regard, are separately regulated.[38] We will return to that issue presently. Platforms used for drilling or the injection of carbon dioxide into the continental shelf do, however, fall under "installations and structures" and are thereby firmly under the jurisdiction of the coastal state.

The continental shelf, the area of primary importance here, comprises the seabed and subsoil of the submarine areas that extend beyond the territorial sea throughout the natural prolongation of a coastal state's land territory to the outer edge of the continental margin,[39] or to a distance of 200 nautical miles from the baselines, where the outer edge of the continental margin does not extend up to that distance.[40] The Baltic Sea is not wide enough for there to be any seabed beyond the continental shelf. Since the continental shelf extends seawards from the territorial sea, the bottom of the EEZ is also the continental shelf of the coastal state.

Without affecting the legal status of the superjacent waters, the coastal state exercises sovereign rights over the continental shelf for the purpose of exploring it and exploiting its natural resources. These rights are exclusive in the sense that even if the coastal state does not explore the shelf or exploit its natural resources, no one else may undertake these activities without the express consent of the coastal state.[41]

Whether the injection of carbon dioxide into the seabed constitutes the utilization of a natural resource, and is thereby subject to the coastal state's sovereign rights, is open to discussion.[42] However, irrespective of whether the coastal state has an exclusive right to store carbon dioxide in the continental shelf, it should have the legal means to exercise control over storage operations in this area. The injection of carbon dioxide is likely to require the establishment and use of installations and structures in the EEZ. As already noted, that is subject to coastal state jurisdiction.[43] In any event, the coastal state has the exclusive right to authorize and regulate drilling on the continental shelf "for all purposes".[44] Even if existing wellbores –

---

[37] Ibid., art. 56(1)(b).

[38] R Lagoni, "Pipelines" in R Wolfrum (ed), *Max Planck Encyclopedia of Public International Law* (e-resource, OUP 2008, updated April 2011) para. 10.

[39] "The continental margin comprises the submerged prolongation of the land mass of the coastal State, and consists of the seabed and subsoil of the shelf, the slope and the rise." UNCLOS, art. 76(3).

[40] UNCLOS, art. 76(1). In fact, the continental shelf may extend even further under certain conditions which, however, are beyond the preview of this analysis since no such areas exist in the Baltic Sea.

[41] UNCLOS, art. 77.

[42] See e.g. Brus, who contends that the legal classification of sub-seabed storage, and thus the preconditions for exercising jurisdiction over such activities, is still very uncertain. M. Brus, "Challenging Complexities of CCS in Public International Law" in M. M. Roggenkamp and E. Woerdman (eds) *Legal Design of Carbon Capture and Storage* (Antwerp: Intersentia 2009) 19–60, at 35.

[43] UNCLOS, art. 56.

[44] UNCLOS, art. 81.

typically from oil or gas exploration – may be used for injecting carbon dioxide, a storage operation could, for practical and safety reasons, not reasonably be initiated without the operator having the right to drill in the seabed. Without prejudging the subsequent discussion on pertinent EU law, it may be noted that the regulation of CCS in EU law is premised on the coastal state having the exclusive right to authorize storage operations.[45]

Regarding pipelines, which can serve a vital role in CCS projects, they are regulated both in the regime for the EEZ and that for the continental shelf, i.e. Parts V and VI of UNCLOS. In practice, however, the provisions regarding the laying of pipelines on the continental shelf are of most significance.[46]

According to Part VI, all states are entitled to lay submarine cables and pipelines on the continental shelf. However, whereas the coastal state may not otherwise impede the laying or maintenance of such cables or pipelines on the continental shelf, it has the right to take "reasonable measures for the exploration of the continental shelf, the exploitation of its natural resources and the prevention, reduction and control of pollution from pipelines".[47] This also means that UNCLOS does not allow for restrictions on the laying or operation of submarine pipelines for other reasons.[48]

The delineation of the course of any pipeline on the continental shelf is also subject to a requirement of consent by the coastal state.[49] This enables such states to establish consent procedures for pipeline laying. As has been discussed elsewhere, the fact that drilling in the seabed is subject to a coastal state consent requirement provides coastal states with added power to regulate pipelines on their continental shelves. The laying of a pipeline is namely preceded by a survey of the seabed, which normally includes drilling.[50]

---

[45] How else could any member state ensure, as it is obliged to, that no storage site is operated without a storage permit in the area to which the Directive applies? Directive 2009/31/EC of the European Parliament and of the Council on the Geological Storage of Carbon Dioxide OJ [2009] L140/114, art. 6. See furthermore, D. Langlet "Transboundary Dimensions of CCS: EU Law Problems and Prospects" (2014) 8 *Carbon and Climate Law Review* 198–207, at 200.

[46] According to UNCLOS Part V on the EEZ the coastal state's sovereign rights and jurisdiction are with respect to the seabed and subsoil to be exercised in accordance with Part VI of UNCLOS, i.e. the rules on the continental shelf. On the relationship between Parts V and VI of the Convention as regards pipelines, see D. Langlet "Transboundary Transit Pipelines: Reflections on the Balancing of Rights and Interests in Light of the Nord Stream Project" (2014) 63 *International & Comparative Law Quarterly* 977–995.

[47] UNCLOS, art. 79(1) and (2). On the notion of "reasonable measures" in this context, see Langlet note 46.

[48] T. Koivurova and I. Pölönen "Transboundary Environmental Impact Assessment in the Case of the Baltic Sea Gas Pipeline" (2010) 25 *International Journal of Marine and Coastal Law* 151–181, at 179.

[49] UNCLOS, art. 79 (3).

[50] See Langlet, note 46 at 988.

Also under the rules for the EEZ all states, and indirectly their citizens,[51] enjoy the right of laying submarine cables and pipelines, but also other internationally lawful uses of the sea, such as those associated with the operation of cables and pipelines.[52]

In summary, coastal states have an unrestricted right to control virtually any CCS activity, be it drilling, the erection of rigs, or the laying of transport pipelines in the territorial sea. The one exception would be the transport of captured carbon dioxide by ship through the territorial sea, as long as the shipment meets the criteria for innocent passage.[53] Furthermore, with respect to the EEZ the coastal state has the legal competence to control most CCS related activities. This is most evident with respect to the construction of rigs and platforms as well as any activity involving drilling.

This control also entails an obligation on the coastal state to exercise its sovereignty or jurisdiction in order to protect the marine environment since states are under a general obligation to take "all measures necessary to ensure that activities under their jurisdiction or control are so conducted as not to cause damage by pollution to other States and their environment".[54] A more specific requirement is that coastal states shall adopt laws and regulations to prevent, reduce and control the pollution of the marine environment arising from or in connection with seabed activities subject to their jurisdiction.[55] With respect to the Baltic Sea there is also the Convention on the Protection of the Marine Environment of the Baltic Sea Area, also known as the Helsinki Convention.[56] It requires its parties, inter alia, to take "all measures necessary to ensure that activities under their jurisdiction or control are so conducted as not to cause damage by pollution to other States and their environment".[57] There are also other, more specific provisions in the Helsinki Convention, to be discussed presently, which have even more immediate implications for the geological storage of carbon dioxide.

Before proceeding to EU law perspectives, it should be emphasized that this responsibility of the coastal state does not abrogate the responsibility of any other state, under whose jurisdiction or control the operator of the CCS related activities is to act to ensure the protection of the marine environment.

---

[51] The freedom pertains to states, not individuals. But in practice the activities covered by the freedom of the high seas are overwhelmingly exercised by private parties. W. Wiese *Grenzüberschreitende Landrohrleitungen und Seeverlegte Rohrleitungen im Völkerrecht* (Duncker & Humblot GmbH 1997) 210.

[52] UNCLOS, art. 58(1).

[53] On innocent passage, see note 31.

[54] UNCLOS, art. 194(2).

[55] UNCLOS, art. 208.

[56] Convention on the Protection of the Marine Environment of the Baltic Sea Area (Helsinki, 9 April 1992) 1507 UNTS 167.

[57] UNCLOS, art. 194(2).

9.4 EU Law Relevant to CCS: General Preconditions

9.4.1 EU Law and Pertinent International Law

The EU, which has "promoting measures at international level to deal with regional or worldwide environmental problems" as an objective of its policy on the environment,[58] is itself party not only to UNCLOS but also to regional marine conventions, including the Helsinki Convention.

Agreements concluded by the Union are binding for EU institutions as well as for its member states.[59] They form an integral part of the EU legal order and even have primacy over secondary EU legislation, i.e. primarily directives and regulations.[60] In principle this means that both the Union itself and the member states are precluded from taking measures that are inconsistent with agreements to which the EU is party. Failure by a member state to ensure compliance with commitments arising from such agreements is a violation of the state's obligations under EU law.[61] However, this does not necessarily translate into full compliance with international agreements, either with respect to the EU or for the individual member states.

Although the Court of Justice of the EU (CJEU) can examine the validity of EU legislation in light of the international agreements by which the EU is bound, it cannot do so "if the nature and the broad logic of the treaty precludes it, or, if the treaty's provisions do not appear, as regards their content, to be unconditional and sufficiently precise".[62] UNCLOS is, according the CJEU, a treaty "the nature and the broad logic" of which prevents its use for examining the validity of EU measures.[63] Furthermore, agreements entered into by the EU with non-member states only have "direct effect" – i.e. can be invoked by individuals and applied by national courts and authorities, without having been (correctly) transposed into national law in the member state at issue – to the extent that they contain a "clear and precise obligation" which does not depend, for its implementation or effects, on the adoption of any subsequent measure.[64]

This means that EU legislation can be inconsistent with e.g. UNCLOS, without there being an affective legal remedy available. The CJEU as well as national courts should interpret EU law in the light of applicable international obligations even if they are not directly applicable.[65] However, such interpretations cannot go beyond a

---

[58] Treaty on the Functioning of the European Union [2008] OJ C115/47 (TFEU), art. 191(1)

[59] TFEU, art. 216

[60] Case C-308/06 *Intertanko and Others* ECLI:EU:C:2008:312, para. 42. On the nature of EU directives and regulation see art. 288 TFEU.

[61] Case 12/86 *Demirel* ECLI:EU:C:1987:400, para.11.

[62] Case C-308/06 *Intertanko*, note 60 at paras. 43–45.

[63] Case C-308/06 *Intertanko*, note 60 at para. 65.

[64] Case 12/86 *Demirel*, note 61 at para.14. On the general precondition of the direct effect of EU law see Case 26–62 *van Gend & Loos* ECLI:EU:C:1963:1 (English special edition).

[65] Case C-308/06 *Intertanko*, note 60 at para 52.

reasonable understanding of the provision that is being interpreted. We will return to the implications of the relationship between EU law and international law when discussing the legal preconditions for CCS activities in the Baltic Sea area. Before that, the legal framework for CCS in EU law will be briefly discussed.

9.4.2   The Regulation of CCS in EU Law

The EU, and indirectly the member states, have adopted one of the most elaborate legal frameworks for CCS in the world, primarily in the form of Directive 2009/31/EC on the geological storage of carbon dioxide.[66] This "CCS Directive", which forms part of EU environmental policy,[67] establishes a legal framework for the environmentally safe geological storage of carbon dioxide to contribute to the fight against climate change.[68] It applies to the geological storage of carbon dioxide in the territory of the member states, their EEZ and on their continental shelves while prohibiting storage outside those areas. The storage of carbon dioxide in the water column, i.e. in the sea itself, is also prohibited. However, storage activities with a total intended storage below 100 kilotonnes, undertaken for research, development or the testing of new products and processes, are exempt from the Directive.[69]

Importantly, it is for each member state to decide whether it will allow the geological storage of carbon dioxide within its territory and, if so, to determine the areas from which storage sites may be selected, pursuant to the requirements of the Directive. The suitability of any specific geological formation for use as a storage site is to be determined pursuant to the criteria set out in an annex to the Directive. If storage is permitted, the procedures for the granting of storage permits must be open to all entities possessing the necessary capacities.[70]

A particular geological formation may only be selected as a storage site, and thus a permit granted, if there is no significant risk of leakage and if no significant environmental or health risks exist under the proposed conditions of use.[71]

---

[66] Directive 2009/31/EC of the European Parliament and of the Council on the geological storage of carbon dioxide ... [2009] OJ L140/114.

[67] That the Directive is based on the Union's environmental policy has implications for the relationship between EU law and the legal orders of the member states. Legal acts based on that policy do not in principle prevent individual member states from taking more stringent protective measures, i.e. measures going beyond the standards set by EU law in terms of environmental and health protection. On the notion of "more stringent protective measures" see Langlet and Mahmoudi, note 25 at 102.

[68] Directive 2009/31/EC, art. 1.

[69] Ibid., art 2.

[70] Ibid., arts. 4(1) and (3) and 6(2).

[71] Ibid., art. 4. A "significant risk" is defined, in art. 3(18), as "a combination of a probability of occurrence of damage and a magnitude of damage that cannot be disregarded without calling into question the purpose of [the] Directive for the storage site concerned". For a critical discussion on this definition, see Langlet and Mahmoudi, note 25 at 272, note 84.

9.5   Legal Premises for CCS in the Baltic Sea Area

It is questionable whether the legislative context in the Baltic Sea area currently is at all consistent with CCS operations. In the following, some of the relevant challenges and obstacles will be discussed, with a particular focus on issues pertaining to the utilization of the seabed.

A first complication is the Helsinki Convention to which all the coastal states of the Baltic Sea are parties. It prohibits, with some exemptions which are not relevant here, dumping in the Baltic Sea. Dumping is defined as "any deliberate disposal at sea or into the seabed of wastes or other matter from ships, other man-made structures at sea or aircraft".[72] Parties may thus not permit the injection of any matter, such as carbon dioxide, into the seabed through any form of installation if it constitutes disposal. The fact that the injection of carbon dioxide into the deep seabed constitutes disposal should be indisputable since the carbon dioxide does not fulfil any function in the seabed, nor is it ever intended to be retrieved therefrom.[73]

Since the EU is itself bound by the Helsinki Convention, its provisions, in principle, take precedence over EU secondary law. It must thus be asked whether legislation that allows for the injection of captured carbon dioxide into the seabed in areas covered by the Helsinki Convention should not in fact be set aside as inconsistent with EU rules of a higher order? Since the pertinent provision of the Helsinki convention is precise and hardly requires any further measures for its application, it seems that the CJEU would be required, if the issue was brought before it, to find that the parts of the CCS Directive that provide for the granting of storage permits cannot apply with respect to the Baltic Sea area.[74]

Since allowing the use of any part of a member state's territory or continental shelf for the purpose of storing carbon dioxide is not compulsory under EU law, there is no risk of member states being forced to choose between violating the Convention, and thereby an EU law obligation, or the CCS Directive, unless the CJEU declares that the latter does not apply in the specific region. Nonetheless, it is clear that the CCS Directive, as well as EU climate change policy, is premised on CCS being a potentially important tool that the member states should not be prevented from making use of – provided that such use is in accordance with the CCS Directive. It is thus quite problematic if member states violate a higher EU norm, i.e. the Convention, if they grant a permit in accordance with the Directive.

---

[72] Helsinki Convention, arts. 2 and 11.

[73] On the interpretation of similar provisions in the 1972 London Dumping Convention and the 1996 London Dumping Protocol, see D. Langlet "Exporting CO2 for Sub-Seabed Storage: The Non-Effective Amendment to the London Dumping Protocol and its Implications" (2015) 30 *International Journal of Marine and Coastal Law* 395–417, at 403.

[74] Arguably, any court or tribunal of a member state against whose decisions there is no judicial remedy under national law must ask the CJEU for a so-called preliminary ruling on the applicability of the CCS Directive – if faced with a case concerning the storage of carbon dioxide in the seabed in an area to which the Helsinki Convention applies. This follows from art. 267 TFEU and Case 283/81 CILFIT ECLI:EU:C:1982:335, paras. 13–16 and 21.

It would also seem that those Parties to the Helsinki Convention that are EU member states are under an obligation, irrespective of whether they have any intention to engage in CCS operations themselves, to work for an amendment of the Helsinki Convention, so as to allow for sub-seabed storage in accordance with the Directive. This should follow from the requirement for loyal cooperation in what is now art. 4 of the Treaty on European Union (TEU).[75] That provision requires member states to facilitate the achievement of the EU's tasks.[76] It has also been interpreted as an obligation on their competent authorities to use any available means to achieve the Union's objectives.[77] Among the EU's tasks is to work for the sustainable development of Europe and aim at a high level of protection and improvement of the quality of the environment.[78] The elaboration of a legal and policy framework to enable and promote CCS operations is evidently part of executing that task.[79]

However, this inconsistency could be overcome also without an amendment of the Convention, e.g. through the adoption by the Parties to the Helsinki Convention of an interpretative resolution indicating that the injection of carbon dioxide – as part of CCS operations carried out in accordance with applicable international law – is not to be considered dumping. That the wording of the dumping prohibition of the Helsinki Convention does not appear to be very ambiguous is not necessarily a problem, as subsequent practice,[80] which an interpretative resolution would constitute, has been found to affect the interpretation of rather clear treaty provisions.[81] However, the adoption of an interpretative resolution requires political will and unanimous support among the parties to the Convention.[82]

Considering the obstacle to storage in the seabed presented by the dumping prohibition in the Helsinki Convention, and also the fact that the geology in parts of the Baltic Sea area does not allow for the geological storage of carbon dioxide, exporting carbon dioxide from the region for storage elsewhere could be an option. However,

---

[75] [2002] OJ C 325/5.

[76] Case C-433/03 *Commission v. Germany* ECLI:EU:C:2005:462, para. 63.

[77] Case C-165/91 *van Munster* ECLI:EU:C:1994:359.

[78] Art. 3(3) TEU.

[79] The CCS Directive establishes a legal framework for the environmentally safe geological storage of carbon dioxide. The purpose of such storage is the permanent containment of carbon dioxide in such a way as to prevent and, where this is not possible, eliminate, as far as possible, negative effects and any risk to the environment and human health, i.e. contribute to the protection and improvement of the quality of the environment. Directive 2009/31/EC, art. 1. On the financial mechanism used to support the development of CCS in the EU, see M. Doppelhammer "The CCS Directive, its Implementation and the Co-financing of CCS and RES Demonstration Projects under the Emissions Trading System (NER 300 Process)" in I. Havercroft, R. Macrory and R. B. Stewart (eds) *Carbon Capture and Storage: Emerging Legal and Regulatory Issues* (Hart, Oxford, 2011) 93–106.

[80] See Art. 31 (3) of the Vienna Convention on the Law of Treaties (Vienna 23 May 1969) (1969) 8 ILM 679.

[81] International Law Commission (ILC) First report on subsequent agreements and subsequent practice in relation to treaty interpretation, by Georg Nolte, Special Rapporteur (19 March 2013) UN Doc. A/CN.4/660, para. 49.

[82] ILC, Second report on subsequent agreements and practice in relation to the interpretation of treaties, by Georg Nolte, Special Rapporteur (26 March 2014) UN Doc. A/CN.4/671, para. 102.

the so-called 1996 London Dumping Protocol,[83] to which Estonia, Denmark, Sweden and Germany are Parties, prohibits the export of wastes or other matter to other countries for dumping.[84] The Protocol's definition of dumping covers any deliberate disposal into the sea of material or substances from vessels, aircraft, platforms or other man-made structures at sea, and any storage of material or substance in the seabed and the subsoil thereof from the mentioned vessels and structures.[85] It is thus clear that the geological storage of carbon dioxide in the sub-seabed constitutes dumping under the Protocol and that parties may not allow the export of carbon dioxide for that purpose. Hence, any arrangements for the exportation of carbon dioxide for storage from Denmark, Estonia, Germany or Sweden could be regarded as a breach of the London Protocol.

In order to enable such export, the parties to the London Protocol adopted an amendment to the relevant provision of the Protocol in 2009. However, the amendment has so far only been accepted by a handful of parties and the prospects for its coming into force look bleak. In principle, this problem could also be overcome, e.g. by the parties' adoption of an interpretative resolution or a subsequent agreement. However, the political will to do so does not seem to exist.[86]

Also in this case, it could be argued that EU member states are obliged, in accordance with the principle of loyal cooperation, to work for the elimination of this obstacle to the extent that it prevents shipments of carbon dioxide for geological storage between member states.[87] But since an amendment to the Protocol requires acceptance by two-thirds of the Parties for it to enter into force, action by EU member states on their own would not solve the problem.[88]

Any export of carbon dioxide for geological storage outside of the combined EU/EFTA area is also precluded by EU waste law since the export of waste for disposal to countries outside that region is prohibited.[89]

Hence, legal obstacles exist to both the storing of captured carbon dioxide in the seabed in the Baltic Sea area and to exporting carbon dioxide for such storage outside of the region, or even within it, to the extent that the state in which the carbon

---

[83] Protocol to the Convention on the Prevention of Marine Pollution by Dumping of Wastes and Other Matter (London, 7 November 1996) (1997) 36 ILM 1 ('London Protocol').

[84] A Working Group set up by the parties to the London Convention found this prohibition to apply to all export of carbon dioxide streams from the jurisdiction of one contracting party to any other country. Report of the 1st Meeting of the Legal and Technical Working Group on Transboundary CO2 Sequestration Issues (3 March 2008) LP/CO2 1/8, para. 1.

[85] Ibid., arts. 4.1 and 4.3.

[86] On the prohibition and the non-effective amendment of the London Dumping Protocol, see Langlet, note 73.

[87] The EU has changed its own waste legislation so as to enable carbon dioxide for geological storage to be transported between member states. Regulation (EC) No 1013/2006 on Shipments of Waste [2006] OJ L190/1 now excludes "shipments of CO_2 for the purposes of geological storage in accordance with Directive 2009/31/EC" from its scope. Ibid., art. 1(3)(h) as amended. See further Langlet, note 45.

[88] London Protocol, art. 21.

[89] Regulation (EC) No 1013/2006 of the European Parliament and of the Council on the shipment of waste [2006] OJ L190/1, art. 34.

dioxide was captured is party to the London Protocol. However, the prospects for overcoming at least the problem posed by the definition of dumping in the Helsinki Convention seem fairly good, if the parties to that Convention who are also EU member states set their minds to achieving the required amendment or clarification.

However, an analysis of the legal preconditions for CCS in the Baltic Sea area need not only engage with law as a potential obstacle to CCS activities. There is also the issue of law as facilitator and arbiter of interests related to the complex and potentially risky activity that any transboundary CCS operation would constitute.

9.6   Conflicting Interests and cooperation Relating to CCS in the Baltic Sea

9.6.1   CCS and Other Legitimate Uses of the Seabed

Clearly, CCS operations, if carried out, have the potential to interfere with other legitimate uses of the sea and the seabed. However, the constructions needed for injecting carbon dioxide into the seabed should only require a fairly small area. And pipelines, once in operation, need not constitute a serious obstacle to other activities such as fishing and shipping. They may interfere with existing or future infrastructure on the seabed, such as other pipelines or cables. There are, however, technical solutions to handle intersections between submarine pipelines and/or cables.[90]

Activities that may be harder to reconcile with pipelines are mining, hydrocarbon exploration, sand extraction and the construction of energy-generating installations, typically wind farms. But, when in place, even large pipelines should not impede such activities more than in their immediate vicinity. Measured along the whole route of the pipeline, this can still add up to a large area. However, for anyone wanting to carry out an activity within a specific region, the obstacles caused by a pipeline passing through that region should not be severe. Nonetheless, the routeing of any CCS-related pipelines should be planned so as to minimize their impact on other legitimate uses of the sea since, as stated in UNCLOS, the exercise of the rights of the coastal state over the continental shelf must not infringe or result in any unjustifiable interference with navigation and other rights and freedoms of other states as provided for in the Convention.[91] This logic is even stronger with respect to the injection points, which may require installations that reach up above the surface

---

[90] See, e.g. the permit for the section of the Nord Stream gas pipelines that traverses the Finnish EEZ/continental shelf, according to which the applicant is obliged to allow crossings using the most financially advantageous construction and to co-operate closely with other operators when planning and realising crossings of gas pipelines and cables. Consent to Exploit Finland's Exclusive Economic Zone (5 November 2009) 678/601/2009 (Unofficial translation provided by the Ministry of Employment and the Economy) 30.

[91] UNCLOS, art. 78(2).

of the sea and thereby have greater potential to interfere with other uses, including navigation. Hence, there is a need for the planning and executing of CCS activities with due consideration for other existing or future legitimate activities in the same area.

Protected areas, particularly those designated as part of the EU's Natura 2000 network of protected areas may conflict with the construction and operation of injection facilities and the laying of pipelines. In the territorial sea as well as the EEZ such activities may, with some narrowly defined exceptions, not be allowed if they can adversely affect the integrity of a Natura 2000 site.[92] The standards set by the CJEU for determining the absence of such a risk are quite demanding.[93]

Furthermore, it is not only conflicts between CCS infrastructure and other legitimate uses that require consideration; the construction and operation of transboundary CCS infrastructure may, in itself, necessitate considerable levels of coordination and the taking into account of plans and activities between the states involved. In order to get such a large, integrated infrastructure to operate safely and even to make it possible to build it without economically or technically insurmountable challenges, there ought to be a high level of coordination in planning, assessment and permit granting.

Obviously, there is also a general need to coordinate measures in crowded marine areas, such as the Baltic Sea, in order to enable the resources of such areas, including the space as such, to be utilized efficiently. All this clearly calls for coordinated planning among the coastal states of the affected region. It is therefore relevant to look at what legal requirements exist in that regard.

9.6.2   International Law

International law in general does not impose very far-reaching or specific obligations on states in terms of cooperation.[94] It would thus be hard to derive from general international law any specific measures that would be required in relation to CCS activities.[95] However, specific obligations pertain to activities that may

---

[92] Directive 92/43/EEC on the conservation of natural habitats and of wild fauna and flora [1992] OJ L 206/7, art. 6. On the significance of Natura 2000 areas for CCS deployment in a regional context, see *CCS in the Skagerrak/Kattegat-region – Final report*, February 2012, available at <http://interreg-oks.eu/webdav/files/gamla-rojektbanken/se/Material/Files/Kattegat/Skagerrak/Dokumenter+projektbank/CCS%20final%20report.pdf> at 75.

[93] On the relevant case law of the CJEU, see Langlet and Mahmoudi, note 25, at 359.

[94] There is of course the fundamental obligation that international disputes are to be settled by peaceful means in such a manner that international peace and security, and justice, are not endangered. Charter of the United Nations (San Francisco, 26 June 1945) 1 UNTS XVI, art. 2(3).

[95] Although the existence of a duty to cooperate may be asserted, that duty "is too indefinite and too vague to require the states to act in a certain specific way or to adopt and implement certain specific measures…". F.X. Perrez *Cooperative Sovereignty: From Independence to Interdependence in the Structure of International Environmental Law* (Kluwer Law International, The Hague, 2000) 261.

significantly affect the environment and even more so with respect to the marine environment.

It is now well established that there is "a requirement under general international law to undertake an environmental impact assessment [EIA] where there is a risk that the proposed industrial activity may have a significant adverse impact in a transboundary context, in particular, on a shared resource".[96] However, the scope and content of such impact assessment is not specified by customary international law. The same seems to be true for the extent to which cooperation is required with other affected States in the carrying out of an EIA and the nature of such cooperation.

With regard to the Baltic Sea, more detailed requirements concerning EIA do, however, follow from the Convention on Environmental Impact Assessment in a Transboundary Context (Espoo Convention), to which all the coastal states except Russia are Parties.[97] The Espoo Convention requires an EIA for large-diameter oil and gas pipelines and major storage facilities for petroleum and chemical products – if they are likely to cause a significant adverse transboundary impact.[98] This may clearly apply to pipelines for transporting carbon dioxide. Whether the provision on storage facilities for chemical products would apply to geological formations under the seafloor that are to be used as reservoirs for carbon dioxide is uncertain.[99] It was hardly an activity considered at the time of the drafting of the Espoo Convention, but a purposive interpretation would suggest that the provision nonetheless applies because leakage could pose an acute threat to life and health. It is also likely that injected carbon dioxide could migrate significant distances, including across borders.[100] Should the Convention not be construed so as to cover geological formations it does provide for treating a non-listed proposed activity as if it were listed. However, the concerned parties then need to agree that the activity is likely to cause a significant adverse transboundary impact in the particular case in question.[101]

With respect to the two Nord Stream pipelines that were constructed in the Baltic Sea in 2010–2012, Russia, which has signed but not ratified the Espoo convention, agreed to apply the Convention to the extent permitted by its national legislation. The states concerned even carried out a more thorough assessment than they were obliged to by the Convention's minimum requirements.[102] However, beyond taking

---

[96] *Pulp Mills on the River Uruguay* (*Argentina v. Uruguay*), Judgment, ICJ Reports 2010, p. 14, para. 204.

[97] Convention on Environmental Impact Assessment in a Transboundary Context (Espoo, 25 February 1991) 1989 UNTS 310 (Espoo Convention).

[98] Ibid., art. 2(3) and Annex 1, points 8 and 16.

[99] R. Purdy "Geological Carbon Dioxide Storage and the Law" in S. Shackley and C. Gough (eds) *Carbon Capture and its Storage: An Integrated Assessment* (Ashgate, Aldershot and Burlington, 2006) 87–139, at 125.

[100] See the "clarifications/assumptions" of the relevant working group under the London Protocol on dumping. Report of the First Meeting of the Legal and Technical Working Group on Transboundary CO_2 Sequestration Issues (3 March 2008) LP/CO2 1/8, 2.

[101] Espoo Convention, art. 2(5).

[102] Koivurova and Pölönen, note 48 at 174.

due account of the outcome of the EIA in the final decision regarding the activity for which the assessment was carried out, the rules on EIA do not provide for cooperation or coordination in the carrying out of such projects.[103]

However, specific cooperation requirements are set out in Part XII of UNCLOS. Indeed, cooperation has been referred to as "a fundamental principle in the prevention of pollution of the marine environment under Part XII of the Convention and general international law".[104] It should thus be clear that cooperation between states is required when needed to prevent marine pollution. In this context, it should be recalled that "pollution of the marine environment" has been given a broad definition in UNCLOS and has even been construed to encompass "measures focused primarily on conservation and the preservation of ecosystems".[105] However, even this does not provide much guidance on how cooperation should be carried out in specific cases. And it does not require cooperation beyond what is needed for environmental protection. It does, e.g. not include any requirement to cooperate in order to ensure that marine natural resources or space at sea are efficiently utilized and that conflicts over such utilization be effectively avoided.

The Baltic Sea qualifies as a so-called semi-enclosed sea as defined by UNCLOS.[106] States bordering such a sea are subject to provisions aiming to regulate cooperation in a wider sense. Such states are expected to cooperate with each other in the exercise of their rights and in the performance of their duties under UNCLOS, e.g., by endeavouring to coordinate the implementation of their rights and duties with respect to the protection and preservation of the marine environment.[107] However, that this provision would entail a binding obligation of any level of substance is disputed.[108] And it is also fairly unspecific in its approach to cooperation.

An often referred to instrument for the coordination of policies with respect to the use of ocean resources and ocean space is maritime (or marine) spatial

---

[103] Parties are also required to, either individually or jointly, take all appropriate and effective measures to prevent, reduce and control significant adverse transboundary environmental impact from proposed activities. Espoo Convention, art. 2(1).

[104] *MOX Plant (Ireland v. United Kingdom),* Provisional Measures, Order of 3 December 2001, ITLOS Reports 2001, p. 95, para. 82.

[105] UNCLOS, art. 1(1)(4) and *Chagos Marine Protected Area Arbitration (Mauritius v. United Kingdom)* Arbitral Award of 18 March 201, para. 538.

[106] The Baltic Sea qualifies as semi-enclosed both because it is "surrounded by two or more States and connected to another sea or the ocean by a narrow outlet" and since it consists "entirely or primarily of the territorial seas and exclusive economic zones of two or more coastal States". UNCLOS, art. 122.

[107] UNCLOS, art. 123.

[108] E. Franckx and M. Benatar "The 'Duty' to Co-Operate for States Bordering Enclosed or Semi-Enclosed Seas" (2013) 31 *Chinese (Taiwan) Yearbook of International Law and Affairs* 66–81. It could be noted, however, that in the Nord Stream case the Finnish Government referred to its participation in the international EIA under the Espoo Convention as its way to dispose of "the obligation to cooperate that pertains to States bordering a semi-enclosed sea like the Baltic Sea". Consent to Exploit Finland's Exclusive Economic Zone, note 90, 23.

planning (MSP).[109] MSP can be understood as "a public process of analyzing and allocating the spatial and temporal distribution of human activities in marine areas to achieve ecological, economic, and social objectives that usually have been specified through a political process".[110]

There is no international legal framework explicitly regulating or prescribing MSP. Even though large parts of UNCLOS and the system of jurisdictional zones that it establishes may be seen as a basic form of MSP, these rules are primarily focused on allocating competences and, as we have seen above, only in rather general terms do they provide for cooperation or the coordination of plans and activities.[111] They do not guarantee an integrated approach to the planning and execution of activities that make use of the sea and the seabed. But a lack of a clear legal obligation is not the same as an absence of action. In the Baltic Sea area, work has been carried out since the 1990s to promote and coordinate MSP within the intergovernmental forum VASAB (Vision and Strategies Around the Baltic Sea).[112] It is guided by a Conference of Ministers responsible for spatial planning and development from the eleven states of the Baltic Sea region. Also HELCOM, the governing body of the Helsinki Convention has emphasized the important role of MSP and cooperates with VASAB in the elaboration of a policy framework for MSP.[113] This has resulted in the adoption of the so-called Baltic Sea Broad-scale Maritime Spatial Planning Principles.[114] According to these, MSP should be developed in a joint pan-Baltic dialogue with coordination and consultation between the Baltic Sea States. Whenever possible, maritime spatial plans should be developed and amended with the Baltic Sea region perspective in mind.[115]

There is hence an active process for promoting and facilitating coordinated MSP in the Baltic Sea area. But there are no guarantees regarding the level of coordination that will be achieved and the development of spatial plans tends to be heavily influenced by national regulatory and planning traditions and frameworks. However, for the states that are EU members, EU law provides a structure for the development

---

[109] On the legal perspectives of MSP in a transboundary context, see D. Hassan, T. Kuokkanen and N. Soininen (eds) *Transboundary Marine Spatial Planning and International Law* (Routledge, New York, 2015).

[110] UNESCO Marine Spatial Planning Initiative at < http://www.unesco-ioc-marinesp.be/>

[111] On the role of UNCLOS for marine spatial planning, see D. Hassan and N Soininen "United Nations Convention on the Law of the Sea as a Framework for Marine Spatial Planning" in Hassan, Kuokkanen and Soininen, note 109, at 60–84.

[112] Regarding VASAB's work with MSP, see < http://www.vasab.org/index.php/maritime-spatial-planning>

[113] On HELCOM's work on MSP, see < http://www.helcom.fi/action-areas/maritime-spatial-planning/>

[114] Adopted by HELCOM HOD 34–2010 and the 54th Meeting of VASAB CSPD/BSR, available at <http://www.helcom.fi/Documents/HELCOM%20at%20work/Groups/MSP/HELCOM-VASAB%20MSP%20Principles.pdf>

[115] Baltic Sea Broad-scale Maritime Spatial Planning Principles, Principle 7, Transnational coordination and consultation.

of marine spatial plans and also for the coordination of such plans. This will be further addressed in the subsequent section after introducing the EU law relevant to CCS.

9.6.3 EU Law

The principle of sincere, or loyal, cooperation is of fundamental importance in EU law.[116] It obliges the Union and the member states to assist each other in carrying out tasks which flow from the Treaties. Member states must take all necessary measures to ensure that their obligations under the treaties and EU secondary legal acts are fulfilled. As previously mentioned, it has also, as regards the member states, been interpreted as an obligation on their competent authorities to use any available means to achieve the Union's objectives.[117] The Commission is obliged to cooperate with member states and not to take measures that prevent them from fulfilling their obligations under EU law.[118] The principle is generally not seen as imposing any more precise obligation on the member states in terms of cooperating among themselves to ensure the achievement of the objectives of EU law. However, secondary EU law, i.e. primarily directives and regulations, does contain obligations of that kind.

Initially it should be recalled that EU law, like international law, has rules on EIA. The EU requirements, set out in Directive 2011/92/EU on the assessment of the effects of certain public and private projects on the environment,[119] are, particularly when read together with the attendant and very extensive case law from the CJEU,[120] more precise than those of both customary international law and of the Espoo Convention. But just as in international law, the EU rules on EIA have little to say about how plans and projects are to be developed and implemented in a transboundary context, and about the level of cooperation or coordination that may be required, even though the EIA must involve other member states as soon as a project is likely to have significant effects on the environment in those states.[121]

An EU Directive that is of particular relevance in this context is obviously the EU's CCS Directive (i.e. Directive 2009/31/EC). Whereas it does not set out a general duty to cooperate it has a provision on "transboundary cooperation" which stipulates that:

---

[116] The principle is now found in art. 4(3) TEU. It has, e.g. been described as the principle "upon which the [EU] Court has built the entire European Union constitutional architecture". A. Dashwood, M. Dougan, B. Rodger, E. Spaventa, and D. Wyatt *Wyatt and Dashwood's European Union Law* (6th edn. Hart, Oxford and Portland, 2011).

[117] Case C-165/91 *van Munster* ECLI:EU:C:1994:359.

[118] Case C-523/04 *Commission v Netherlands* ECLI:EU:C:2007:244, para. 34.

[119] [2012] OJ L26/1.

[120] For an introduction to this case law, see Langlet and Mahmoudi, note 25, section 7.1.

[121] Directive 2011/92/EU, art. 8a.

In cases of transboundary transport of CO_2, transboundary storage sites or transboundary storage complexes, the competent authorities of the Member States concerned shall jointly meet the requirements of this Directive and of other relevant Community legislation.[122]

However, the provision does not provide any guidance as to how, for example, conflicting technical requirements or competing interests are to be handled.

Some level of coordination could be achieved through the fact that the EU Commission has a right to see and comment on draft storage permits under the CCS Directive. But the only actual obligation that follows is that the member state concerned must state its reasons if it departs from the Commission's opinion.[123] Also, these comments will come at a rather late stage in the process when the prospective operator has already invested considerably in the development of a permit application for a particular site. To truly coordinate CCS policy for the benefit of the effective and efficient utilization of resources, such as geological storage space and the necessary infrastructure, additional initiatives are likely required between neighboring states with an interest in CCS.

Particularly with respect to potentially competing interests in the seabed, MSP is broader in scope and more promising in terms of the coordination of policies and the avoidance of conflict. After some less than successful attempts to promote integrated coastal zone management, the EU adopted Directive 2014/89/EU in 2014, establishing a framework for maritime spatial planning (MSPFD).[124] For the purpose of the Directive, "maritime spatial planning" is a process by which the relevant member state's authorities analyze and organize human activities in marine areas to achieve ecological, economic and social objectives. These activities shall aim to contribute to the sustainable development of energy sectors at sea and of maritime transport as well as to the protection of the environment.[125]

Although most of the MSPFD's requirements are procedural rather than substantive, it requires member states to set up maritime spatial plans identifying the spatial and temporal distribution of relevant existing and future activities and uses.[126] The plans are to take into consideration the relevant interaction of activities and uses. Furthermore, the development of the plans must ensure transboundary cooperation between member states and promote cooperation with third countries.[127] As part of the planning and management process, member states bordering marine waters shall cooperate with the aim of ensuring that maritime spatial plans are coherent and coordinated across the marine region concerned. In particular, issues of a transnational nature must be taken into account. Cooperation shall also be endeavored with third countries on their actions with regard to maritime spatial planning in the

---

[122] Directive 2009/31/EC, art. 24.

[123] Ibid., art. 10.

[124] [2014] OJ L257/135.

[125] Ibid., arts. 3(2) and 5(2).

[126] Ibid., art. 8.

[127] Ibid., art. 6.

relevant marine regions.[128] The cooperation is to be pursued, inter alia, through existing regional institutional cooperation structures, such as HELCOM, or through the networks or structures of the member states' competent authorities.[129]

To the greatest extent possible the implementation of the MSPFD is to build upon existing national, regional and local rules and mechanisms.[130] The Directive is generally deferent to national policies and priorities and rather than introducing new sectorial policy targets which the plans should reflect, it aims to integrate and link the objectives defined by national or regional sectorial policies.[131] The proposal for a directive on maritime spatial planning initially tabled by the EU Commission was more strongly worded with respect to both cooperation among member states and cooperation with third countries.[132]

Whereas the Commission's initial proposal included a list of activities that have to be taken into consideration when establishing maritime spatial plans, the Directive as adopted only provides an indicative list of activities. Neither mentions CCS activities specifically.[133] The Directive does list installations and infrastructures for the production of energy from renewable sources, i.e. other offshore climate change mitigation technologies, such as wind farms and wave and tidal energy plants. It also lists submarine cable and pipeline routes. And states can obviously use maritime spatial plans as well as the processes for their elaboration to cooperate on other aspects of CCS.

Despite its shortcomings, the MSPFD is probably the most promising legal instrument for promoting and facilitating a coordinated approach to seabed uses, including CCS activities. However, much will depend on the quality of the regional processes needed for coordinating plans and the extent to which the plans will deal with CCS activities in a sufficiently precise manner, so as to provide clear guidance for authorities and prospective operators. This remains highly uncertain considering the overall dire state of CCS in the EU. Nevertheless, irrespective of the future prospects for CCS, important lessons can be drawn from studying the preconditions for its employment in the Baltic Sea area as well as the increased need for the required cooperation and coordination of plans and policies.

---

[128] Ibid., art 12.

[129] Ibid., art 11.

[130] Ibid., preambular para. 12.

[131] *Proposal for a Directive of the European Parliament and of the Council establishing a framework for maritime spatial planning and integrated coastal management* (12 March 2013) COM(2013)133final. 5.

[132] Compare ibid. arts. 12 and 13 and Directive 2014/89/EU, arts. 11 and 12. For an analysis of the differences between the proposal and the Directive as eventually adopted, see A. Zervaki "The Legalization of Maritime Spatial Planning in the European Union and Its Implications for Maritime Governance" (2016) 30 Ocean Yearbook 32–52.

[133] COM(2013) 133 final, note 131, art. 7; and Directive 2014/89/EU, art. 8.

9.7 Conclusions

Of all climate change mitigation technologies making use of the continental shelf, CCS offshore storage of carbon dioxide is likely the most legally complex one and probably also the one that affects most other interests. Whether CCS will ever be employed in the Baltic Sea area is uncertain, even though it is often referred to as an important option, particularly for industries that generate large quantities of carbon dioxide in their production processes. However, since CCS is a complex activity that is likely to involve offshore installations, submarine pipelines and drilling in the continental shelf, an analysis of the legal preconditions for CCS in the Baltic Sea area inevitably covers many issues that have relevance for other climate change mitigation technologies such as offshore installations for renewable energy generation. There are also issues that are particular to CSS and which reveal significant inconsistencies between the regulatory frameworks applicable at international and EU level.

Overall, UNCLOS provides a good basis for regulating CCS operations through a fairly clear allocation of competencies and attendant responsibilities with respect to the continental shelf and the EEZ. A coastal state is equipped with the requisite legal authority to control the use of its continental shelf for the geological storage of carbon dioxide. Submarine pipelines are differently regulated than other aspects of offshore CCS but also in this case the coastal state will be able to exercise significant or even controlling influence over the relevant activities. This also means that the coastal state will, in many respects, be the arbiter between the competing interests pertaining to CCS activities. It will also have an obligation to ensure that the marine environment is sufficiently protected.

Other pieces of law represent significant obstacles to CCS activities. Some of these are regional in scope whereas others are potentially global. The most significant provision in this respect is probably the prohibition on dumping in the Helsinki Convention which, due to the definition of dumping, precludes the injection of carbon dioxide into the seabed in the Baltic Sea. This potentially negates the Union's CCS Directive by prohibiting – under EU law due to the Status of the Helsinki Convention in the Union's legal order – the storage of carbon dioxide in the Baltic Sea. There are also restrictions in both international and EU law on the export of captured carbon dioxide for geological storage. However, some of these obstacles, including the one posed by the Helsinki Convention, should not be too difficult to overcome although it would require coordinated action by the affected EU member states.

Due to the complexity of CCS operations, but also in order to ensure that the scarce resources of the sea, including space, are used efficiently and without prompting unnecessary conflicts, a significant level of cooperation and the coordination of plans and activities are likely required. This is particularly pertinent with respect to a small and heavily utilized sea area like the Baltic Sea. International law requires environmental impact assessment in certain cases but provides little guidance as regards cooperation on how specific activities are to be planned and executed and does not include assessment beyond environmental protection. There is, however, ongoing work among the Baltic Sea states to promote a coordinated and integrated approach to the planning and utilization of ocean space, although not in a legally binding form. In this respect EU law, and in particular the framework directive on maritime spatial planning has greater potential. However, this legislative framework pays great deference to national interests and priorities, thus it remains to be seen how far planning, not to mention the actual implementation of activities, will be effectively coordinated. And to what extent such coordination will consider the possible deployment of CCS in the region.

Chapter 10
Concluding Remarks: Regulatory Gaps and Broader Governance Patterns in the Baltic Sea

Henrik Ringbom and Marko Joas

Abstract The concluding chapter summarizes the findings on the book and makes a series of conclusions relating to jurisdictional matters as well as the substantive topics covered. It is noted that several regulatory gaps still exist in the regulation of the environment in the Baltic Sea and seeks to use the sample of issues discussed in the book for making some more general observations on the development, nature and consequences of various types of gaps in the specific Baltic Sea context.

Keywords UNCLOS · Regulatory gaps · Multi-level governance · Regulatory layers · Baltic Sea · Policy steering mechanisms

10.1 Introduction

The preceding chapters have discussed specific topics of relevance on the question of how the Baltic Sea is governed, particularly focusing on regulation and legislative gaps and uncertainties. The chapters illustrate the richness of legal issues that are currently relevant in the region within a broad variety of maritime topics. Although the chapters cover their own distinct fields and issues that are not at first sight related to each other, they all cover areas where the legal situation is complex and uncertain, exposing regulatory gaps.

The selection of themes for this book highlights various legal issues of current relevance for the Baltic Sea, giving rise to interesting questions relating to how laws at different levels interact with each other and with other policy tools. The substantive

H. Ringbom (✉)
Department of Law, Åbo Akademi University, Turku/Åbo, Finland
e-mail: henrik.ringbom@abo.fi

M. Joas
Public Administration, Åbo Akademi University, Turku, Finland

© Springer International Publishing AG, part of Springer Nature 2018
H. Ringbom (ed.), *Regulatory Gaps in Baltic Sea Governance*, MARE
Publication Series 18, https://doi.org/10.1007/978-3-319-75070-5_10

focus lies on rules that relate to the protection and preservation of the marine environment, not least because this field of law-making is the most developed internationally in the region. However, the selection is neither complete nor systematic, nor is it necessarily even representative of current policy priorities in the region. Nevertheless, the chapters include at least certain aspects of some of the key regulatory challenges in the region and therefore, despite their limitations in terms of representativeness, provide a basis for some observations of a more general nature on the governance of the Baltic Sea.

The regulatory backdrop for the issues considered is the relatively recent enclosing - in a jurisdictional sense - of the Baltic Sea. As Erik Franckx describes in his chapter, essentially every part of the sea is now covered by zones that belong to the jurisdiction or sovereign rights of one of the nine coastal states and the precise limits of the boundaries are, with a few exceptions, settled. The potential impact of this enclosure can be expected to be largest in those fields where the jurisdiction of states is most comprehensive, including the regulation of the seabed and of the protection and preservation of the marine environment.

However, even a very cursory reading of the different chapters suffices to confirm that the clarification of the jurisdictional borders does not necessarily guarantee clarity or the absence of gaps concerning individual regulatory issues.

The gaps and challenges discussed in the individual chapters differ in their origin, scope, consequences and significance. Similarly, the mechanisms to fill the legal gaps vary from one substantive area to another. While many uncertainties at the level of international regulation may be resolved by traditional legal mechanisms, such as various forms of treaty interpretation, 'soft law'complements, or by the more specific implementing of legislation at national or subnational level, other gaps may require other forms of governance mechanisms.

10.2   Jurisdictional Rules (UNCLOS)

The very stable jurisdictional regime that currently prevails in global ocean governance provides a general law of the sea backdrop for all contributions to this book. This circumstance, which is far from self-evident in a historical perspective, derives from the broad and largely undisputed authority of UNCLOS together with its comprehensive approach with respect to all uses of the seas. The 'Constitution for the Oceans' is today very widely accepted in formal terms, including by all coastal states in the Baltic Sea and its catchment area,[1] and its authority as a principal regulatory starting point for the whole breadth of ocean usages is not in doubt. This provides unprecedented stability for the basic jurisdictional setting, i.e. in relation to what states can or cannot do with respect to different sea areas, which activities are permitted in different places and which are not as well as how to handle

---

[1] See   e.g.   <www.un.org/Depts/los/reference_files/chronological_lists_of_ratifications. htm#The%20United%20Nations%20Convention%20on%20the%20Law%20of%20the%20Sea>

disagreements on the division and use of ocean spaces. In a region that is traditionally marked by political differences and major ideological divisions, this starting point should not be underestimated as a basis for stability through more detailed regulation, given that jurisdictional and political stability support each other.

In the Baltic Sea, moreover, jurisdictional stability is further emphasized by the fact highlighted in the early chapters of the book, that the Baltic Sea is by now 'closed' in a jurisdictional sense. The Baltic Sea no longer includes areas beyond the jurisdiction of any state and any part of it, whether marine areas or parts of the seabed, will thus belong to a given coastal state that exercises jurisdiction or sovereign rights over the area and its resources. This completeness of the Baltic Sea in jurisdictional terms does not mean that the littoral states are free to regulate their maritime zones as they wish; a series of limitations follow from UNCLOS in respect of other states' rights to navigate, lay pipelines and undertake scientific research and generally their duty to have due regard to the interests of other states' interests. However, for all issues discussed in the book, the current jurisdictional setting of the Baltic Sea entails that the entire sea area is covered by areas over which coastal states exercise sovereignty or 'sovereign rights' and hence hold significant jurisdiction on regulating activities and taking associated enforcement measures.

Jurisdictional completeness also implies that there is always a formal body – the coastal state –in charge of the overall management of a particular sea area, e.g. as regards the prioritising and handling of conflicts between different uses of the area. A regulatory gap at the international level may accordingly be addressed by legislation or other guidance at national or sub-national level, within the boundaries of the jurisdiction. Moreover, the jurisdictional enclosure emphasizes that the governance of the Baltic Sea is up to the littoral states alone; there are no more high seas segments or other areas beyond national jurisdiction to provide an 'excuse' for regulatory imperfection. While coastal states' jurisdiction is generally facultative, in the sense that states are under no obligation to make use of it, a number of environmental provisions of UNCLOS impose duties on states to protect and preserve the marine environment.[2] The jurisdictional enclosure of the Baltic Sea has thus also enlarged the scope of the coastal states' environmental duties in the region.

While such an authoritative and unusually settled jurisdictional setup provides a stable legal framework for states to operate within the Baltic Sea, it does not suggest that jurisdictional uncertainties on how the sea may be used have disappeared. First, there are issues that fall beyond the scope of UNCLOS. Environmental concerns that have arisen since UNCLOS was negotiated, and hence were not known at the time, may serve as examples. The increasing concerns related to climate change and ocean acidification, for example, touch upon matters that are simply not addressed in UNCLOS. For such matters, the preamble merely states that such issues are to be

---

[2] Apart from the general duty to protect the marine environment which features in UNCLOS arts. 192, 194 and 195, there are specific obligation with respect to areas under coastal states' authority, separately for different sources of marine pollution (e.g. arts. 207, 208, 210 and 211) some more general cooperation duties, including arts. 123 and 197.

governed by 'general rules of international law', but the role and responsibilities of states and various international institutions remains unsettled.

Second, issues that took place before the conclusion of UNCLOS will not be assessed under its terms. A case in point are the large quantities of ammunition and other toxic materials dumped in a number of places in the Baltic Sea.[3] These materials were dumped following the two World Wars, before any international prohibitions were in place in areas that at the time constituted the high seas and by states that might not exist anymore or where the principles of state succession are not altogether obvious.[4] On the one hand, there is the principle of non-retroactivity that cautions against the application of current international standards regarding jurisdiction or state responsibility when seeking to identify the present day responsibility of the dumping state. On the other hand, however, the mere presence of such materials on the continental shelf may in certain situations create duties for the coastal states to take active measures to have them removed.[5] That removal operation could in turn give rise to international responsibility for the coastal state, e.g. for failing to prevent the further pollution of the marine environment.[6] The legal situation, in other words, suggests that the legal risks are larger for states that actively seek to remove an identified environmental hazard than for those that turn a blind eye to the matter. Such regulatory anomalies in international law may be counterproductive from an environmental perspective and hence generate tension with other rules of environmental law. The situation calls for close collaboration and active policy measures by the states involved, including the state that originally dumped the matter, where identifiable.[7]

Third, the mere presence of an authoritative legal system and laws does not tell us much about the quality or content of the system or laws. Several chapters in this book illustrate how key questions on individual subject matters relate to how the

---

[3] See e.g. H-J Heintze Legal problems related to old chemical munitions dumped in the Baltic Sea, in T. Stock, & K Lohs (eds), *The challenge of old chemical munitions and toxic armament wastes* (SIPRI Chemical and Biological Warfare Studies, Stockholm 1997) 255–262. On the extent of the issue, see also HELCOM Doc. *Chemical Munitions Dumped in the Baltic Sea. Report of the ad hoc Expert Group to Update and Review the Existing Information on Dumped Chemical Munitions in the Baltic Sea (HELCOM MUNI) 2013*, available at www.helcom.fi/Documents/Ministerial2013/Associated%20documents/Background/Dumped%20chemical%20munitions%20in%20the%20 Baltic%20Sea.pdf and www.chemsea.eu

[4] See e.g. M. Koskenniemi and M. Lehto 'La succession d'Etats dans l'ex-URSS, en ce qui concern particulièrement les relations avec la Finlande' (1992) 38 *Annuaire Français de Droit International* 179, A.M. Beato 'Newly Independent and Separating States' Succession to Treaties Considerations on the Hybrid Dependency of the Republics of the Former Soviet Union' (1994) 9(2) *The American University International Law Review* 525.

[5] Notably UNCLOS art. 194(1) and (3).

[6] See e.g. UNCLOS arts. 194(2), 195 and 198.

[7] See also Report of the Secretary-General, Cooperative Measures to Assess and Increase the Awareness of Environmental Effects Related to Waste Originating from Chemical Munitions Dumped at Sea, UN Doc. A/68/258 (2013). In a Baltic Sea setting, the matter has been addressed within the HELCOM expert working group on environmental risks of hazardous submerged objects (http://helcom.fi/helcom-at-work/groups/response/submerged)

jurisdictional rules are to be understood and interpreted. In some cases, as is the case on the interpretation of the status of international straits in the Baltic Sea (chapter by Pirjo Kleemola-Juntunen), the interpretations may be steered by political or strategic interests. In other examples the need for interpretations is provoked by open-ended provisions in UNCLOS (examples in this book may be found regarding transnational pipelines or historical wrecks, as discussed by Peter Wetterstein and Jan Aminoff) or arise from new issues that were not foreseen at the time of its adoption and therefore only partially regulated therein (carbon storage underneath the continental shelf, as discussed by David Langlet).

Finally, state practice, including regulatory developments, have increasingly begun to deviate from the system laid down in UNCLOS, not least in the field of marine environmental protection. In certain areas, the jurisdictional framework provided by the law of the sea is therefore increasingly out-dated in view of the developments and challenges of environmental law in practice in the past few decades. Modern regulatory tools strive to move away from the compartmentalisation which is inherent in the law of the sea. While UNCLOS tightly delineates the jurisdictional rights and duties, in geographical terms (through the division of sea areas into various artificial zones) and also substantive terms (through a differentiation different specified activities, such as navigation, fisheries, exploitation of non-living resources etc.), modern environmental governance and law tend to address matters in an integrated (multi-sector) manner, addressing cumulative threats and taking departure in the needs of eco-systems rather than the ocean users.[8] Eco-systems have different needs and therefore require measures that differ from one area to another. Among the chapters in this book, the clearest example of this trend is the study by Brita Bohman on the regulatory framework dealing with the greatest regional environmental concern of the Baltic Sea, i.e. eutrophication.

10.3 Substantive Rules

Turning to the substantive issues discussed in the book, different scenarios and challenges apply for the different themes. With respect to regulatory gaps, the chapters cover three main scenarios.

The first scenario relates to complete regulatory voids. Such voids are not so common anymore, following several decades of intense environmental rule-making at all regulatory levels. However, as the chapter by Lena Gipperth and Thomas Backhaus illustrates, there are still important issues that fall in between current international legal requirements. Chemical mixtures represent an issue that has

---

[8]The most topical example of this shift in the law of the sea is presented by the current negotiations to develop an international legally binding instrument under UNCLOS on the conservation and sustainable use of marine biological diversity of areas beyond national jurisdiction, based on UN General Assembly resolution 69/292 of 19 June 2015. See www.un.org/depts/los/biodiversity/prepcom.htm

escaped the attention of current chemical regulation, partly because of the complexities involved in its regulation and partly because of the difficulties in adopting and enforcing any regulation that targets such mixes and 'chemical cocktails'. Even modern regulatory efforts, mainly by the EU, have failed to regulate the matter in an effective way. The study by Gipperth and Backhaus illustrates how other forms of – less targeted – EU rules have gradually entered this regulatory space and what their impact has been.

The second scenario is where the issue in question represents an environmental concern in only certain parts of the world and has hence not been subject to regulatory attention at the global level. The most important example for the Baltic Sea is the regulation of eutrophication, which is discussed in the chapter by Brita Bohman. As increased levels of nutrients in the sea is not of major concern at global or even European level, the Baltic Sea states have limited regulatory support at global, or even EU level, for addressing this question. As Bohman illustrates, a variety of legal tools of varying normative intensity have been introduced to address this concern from a specific Baltic Sea point of view. Yet the implementation of these tools varies significantly between states.

For both scenarios, a new approach in environmental rule-making over the past decade may offer a solution for addressing regional gaps or specific features at local level in the future. The tendency towards more flexible ecosystem-oriented rules is notable both globally and at EU level. Such rules typically outline the main objectives to be achieved in broad environmental terms, but leaves individual states and regions large amount of discretion for defining and implementing the necessary measures, depending on their particular (ecological) needs.[9] Key environmental legislation at the EU level, such as the Water Framework Directive and the Marine Strategy Framework Directive, are examples discussed in the book. Both are complemented by various kinds of 'soft law'instruments and occasionally further clarified in case law.[10] However, regulatory discretion also implies regulatory uncertainty, which in turn tends to "filter downwards" in the regulatory system.[11] The more concrete measures for defining and implementing the requirements will often be transferred to the national or subnational level. In addition, as the example of the

---

[9] See e.g. S. Söderström & K. Kern 'The Ecosystem Approach to Management in Marine Environmental Governance: Institutional Interplay in the Baltic Sea Region', *Environmental Policy and Governance*, August 2017 DOI: https://doi.org/10.1002/eet.1775

[10] See e.g. the clarifications of the extent of the obligations of the Water Framework Directive (Directive 2000/60), which were made by the Court of Justice of the EU in the *River Weser* judgment, Case C-461/13, *Bund für Umwelt und Naturschutz Deutschland e.V. v. Bundesrepublik Deutschland.*

[11] As has been observed by Bohman & Langlet, "the ecosystem approach is not only hard to define with precision but might also get into conflict with traditional legal requirements for legal certainty and coherence. Such conflicts are seldom addressed in substance in international or EU law, but rather left to individual states to manage in the concrete circumstances." B. Bohman & D. Langlet 'Floater or Sinker for Europe's Seas? The Role of Law in Marine Governance', in M Gilek & Kristine Kern (eds) *Governing Europe's Marine Environment: Europeanization of Regional Seas or Regionalization of EU Policies?* (Ashgate Publishing, 2015) at 70.

governance of eutrophication in the Baltic Sea illustrates, much of the supporting activity is undertaken outside traditional governmental regulatory frameworks and includes a series of network-based collaboration frameworks through information-sharing as well as economic steering instruments at sub-national levels of governance, including NGOs, as main actors, but also involving traditional intergovernmental organisations such as the EU.

The third and probably most common scenario is that rules exist at international and/or EU level for addressing the matter at hand, at least in general terms, but that a closer analysis of a specific substantive issue reveals insufficiencies in terms of coverage and/or compatibility between the rules. This scenario more or less applies to the remaining chapters of the book, focusing on shipwrecks and seabed activities, through issues that are of particular concern to the Baltic Sea region.

The issue of historic wrecks is a matter of particular interest in the Baltic Sea, due to the favourable natural conditions for preserving such wrecks. The matter is dealt with inconclusively in jurisdictional terms in UNCLOS. In order to supplement this regime and provide better protection for historic wrecks, a new special global convention has been developed under the framework of UNESCO. However, as Jan Aminoff illustrates in his article, only one state in the Baltic Sea region has ratified the 2001 UCH Convention and it does not therefore apply in the region. In the absence of such rules, the matter is governed by the international law of salvage, which was not primarily developed to deal with historic wrecks. International legal gaps are filled by national rules, but in this example, even the applicable national rules (including the Antiquities Act, Lost Property Act) are not mainly drafted with wrecks in mind, and their applicability depends on whether or not the wreck is located in the Finnish territorial sea. Since the situation is similar in other littoral states a gap accordingly exists for historic wrecks in the parts of the Baltic Sea that extend beyond the 12 nautical miles limit of the territorial seas. For non-historic wrecks the situation is equally complex with respect to issues that are not addressed in the international conventions, such as ownership or abandonment of ships. As Aminoff's article illustrates, national rules create a complex web of rules, partly based on general civil law, partly on maritime law, little of which has been specifically designed to deal with the regulation of wrecks. Some further clarity in the field is expected over time due to the increased applicability of another successful international harmonization effort, the Nairobi Convention on Wreck Removal, which is now being implemented by the Baltic Sea states.

Another study relating to wrecks is the one Markku Suksi who analyses the legal framework for the removal of wrecks and the availability of options for (Finnish) public authorities to demand action in this regard. Traditionally, in the absence of international rules in this field, states have chosen different routes to deal with the matter. In Finland several responses have been available and different legal regimes have applied depending on whether the wreck was seen as an environmental threat, an obstacle for navigation or just waste or even litter. Suksi's study illustrates how the introduction and implementation of a new international convention, which technically does not even apply in the territorial sea of its states parties, has increased clarity in Finnish legislation to the responses that public authorities have at their disposal with respect to wrecks in Finnish waters. Yet, the availability of alternative legal routes and tools for enforcing wreck removal operations still involves uncertainty in the scope of the obligations for both public authorities and ship operators.

By contrast, the regulations governing international pipelines are reasonably clear from a jurisdictional point of view. UNCLOS clarifies that this is a freedom that extends to the continental shelves of other states, but that the coastal state retains the right to take reasonable measures to protect the environment and certain other interests from cables or pipelines.[12] UNCLOS treats the matter in terms of a bilateral balancing of the interests between the state that wishes to lay the pipeline and the shelf state. A similarly bilateral approach applies to related rules, such as the Espoo Convention on environmental impact assessment. However, inherently international pipeline projects such as the Nord Stream gas pipeline, addressed by Peter Wetterstein, illustrates the complications linked to a purely bilateral approach, involving differing rules, different procedures and related complications for all parties, including the cable laying company. A particularly interesting aspect of this 'bilateralist' approach to international energy projects is that the civil liability of damage caused by the pipeline will eventually depend on the national rules of the shelf state concerned and hence vary depending on where the damage originates and/or occurs. In the absence of international liability and compensation rules, for pipelines and indeed for the whole offshore energy sector, national rules will fill the voids and hence result in significant differences which are to some extent moderated through private international law. While Wetterstein does not see a need for a global instrument in this area, he proposes that EU law might represent the right level to fill the regulatory gaps in the longer run.

The final case study addressed in the book examines how a new issue that was entirely unknown a few decades ago has been settled by regulators, globally and specifically for the Baltic Sea. The concerns related to the sub-seabed storage of carbon dioxide were not – and could not have been – foreseen by the drafters of UNCLOS and the matter is accordingly not discussed in detail in the convention or other instruments that could be of potential use. Nevertheless, quite a few regulatory developments have taken place in recent decades at the international, regional and EU-level. David Langlet guides us through the relevant rules and illustrates that – far from being a regulatory void – the matter is already governed by a series of rules of varying degree of specificity, in this case with the EU rules represent the most specific layer. Nevertheless, the regulatory layers of rules are not always compatible with each other when it comes to their application for this activity in the Baltic Sea and there is little legal guidance on coordination between states involved in, or affected by, sea-based carbon storage. Similarly, coordination between the different (conflicting) regulatory layers is missing. Whereas the prohibition of the geological storage of carbon dioxide in the seabed, which previously was included in the 1996 London Protocol, has been removed, a very similar provision in the Helsinki Convention has not been addressed, which effectively rules out such storage in the Baltic Sea area. Since the EU is itself a contracting party to the Helsinki Convention

---

[12] UNCLOS art. 79.

this prohibition is also part of EU law, thereby rendering a key part of the EU legislation pertaining to carbon capture and storage virtually inapplicable in the Baltic Sea region. This is not merely a systemic inconsistency but also a very real impediment to transboundary storage operations in the area and thus something that ought to spur action by the Union and the member states concerned. The case study also illustrates how actions at global level, in the form of the export prohibition imposed by the London Protocol, can create an obstacle to regional development in the Baltic Sea. The resolution for the carbon storage in the Baltic Sea depends on acceptance of the amendment to the London Protocol, which is a matter that its (global) membership may not share views on or priorities with the Baltic Sea coastal states. Accordingly, even before any carbon storage activity has taken place in the Baltic Sea, it is clear that the legal situation is in need of further clarification.

10.4   General Notes on Environmental Regulation

The main focus of the themes covered in this book is on - broadly defined - the environmental protection of the Baltic Sea. This focus may, as a starting point, be assumed to cater for a relatively complete or 'gap-free' international regulatory framework. Not only is the Baltic Sea among the most heavily regulated sea areas of the world, but the subject area entails several features which stimulate international regulation. Most importantly, the topic by its nature requires international action if it is to be effective. The inter-dependency caused by the trans-border impacts of marine pollution and resource exploitation are easily-understood and further accentuated in a sensitive ecosystem such as a semi-enclosed sea with limited volumes and exchange of water. Apart from that, rules aimed at protecting and preserving the marine environment do not - as opposed to e.g. military security - touch upon core aspects of statehood or sovereignty. It has therefore been politically acceptable for states to assume binding international obligations in this field and there is already a comparatively long tradition of international regulation of the marine environment, both globally and regionally in the Baltic Sea. More recently, the Baltic Sea has been 'enclosed' in terms of maritime zones, meaning that any part of it is subject to one of the coastal states 'sovereignty or sovereign rights. Finally, through the successive enlargement of the EU, the regulatory environment of the Baltic Sea has also become increasingly affected by the growing body of EU environmental legislation.

Nevertheless, as has been illustrated in this book, there are still quite a few gaps and uncertainties regarding regulation. Only a few issues have been highlighted in the book and the small sample is by no means sufficient for drawing general conclusions on the state of the regulation and regulatory gaps in the Baltic Sea. Yet, the sample illustrates the variety and complexity of different kinds of regulatory gaps and calls for some general observations.

10.5   Gaps and Regulatory Layers

There are no *a priori* preferences for a certain legal 'layer' to deal with a particular subject matter. International law is horizontal in nature and does not indicate hierarchical priorities for global rules over regional ones, for example. Nor is there a division of roles between the global and regional rules in substantive terms in the law of the sea, apart from shipping, where a strong preference for global rules is laid down in UNCLOS. There is accordingly little guidance as to what regulatory layer, if any, would be the right one to address a given issue of concern regarding the Baltic Sea. Nothing precludes that multiple layers operate simultaneously and there are numerous examples of that in this book. EU rules that implement or merely duplicate international rules will significantly strengthen the legal framework around them, not only in terms of enforcement powers against non-complying member states and individuals, but also in terms of strengthening the hierarchical position of such international rules.[13] On the other hand, multiple regulatory layers generate obvious risks for confusion, inconsistencies and even conflicts.

The examples in this book highlight the importance of international (global) rules for avoiding regulatory gaps. Without global rules, other layers appear less likely to take regulatory action at regional level. This is particularly the case with respect to public international law instruments of a regional scope, i.e. the Helsinki Convention, which has generally been careful to closely follow - and cautious not to exceed - the requirements of its global counterparts. Where regional requirements exceed the international ones, sacrifices tend to be made in normative strength, meaning that environmental goals that apply only in the Baltic Sea usually take the form of 'soft law'instruments which are difficult to enforce. Regulation at the EU level is somewhat more independent from international rules and does come with significant normative strength, but will rarely be tailor-made for the problems facing the Baltic Sea and will always involve a geographical gap in coverage for the Baltic Sea because one key littoral state is not covered by EU-rules.

It follows from the small sample of issues discussed in this book that the number of regulatory layers involved in addressing any particular matter is not a good standard for assessing how well the regulation tackles the substantive issue at stake. As the case study on carbon storage illustrates, the involvement of three different international regulatory layers (global, regional and EU rules) easily generates confusion or even conflicting rules, unless perfectly coordinated. Perfect coordination, on the other hand, is unlikely to be the norm - in regulation as well as in other governance instruments - in view of the variations in membership, priorities and procedures between the responsible institutions.

---

[13] Since the turn of the Millennium, the Court of Justice of the EU has repeatedly confirmed that international agreements to which the Union is a party, even if only alongside its member states, are part of the EU legal order and are hierarchically above the Union's own secondary legislation, i.e. directives and regulations. See e.g. Cases C-344/04 *IATA and ELFAA*, para. 35 and C-308/06 *Intertanko and others*, para. 42.

10.6   Substantive Gaps

As regards substantive areas covered by this book, it was already noted there are relatively few gaps in the Baltic Sea in the field of jurisdiction. The enclosure of the Baltic Sea in terms of maritime zones and the general robustness of the law of the sea as provided by UNCLOS means that there are few jurisdictional voids in the region. Essentially such open voids only exist in areas that are not regulated in UNCLOS at all. However, as was also noted, that development does not mean that there are no disputed issues any more regarding the law of the sea in the Baltic Sea region. Rather, as the text by Kleemola-Juntunen on passage rights through straits illustrates, it only means that the scope of the gaps has been reduced to cover (the numerous) issues where the interpretation of the jurisdictional rules may present significant divergences.

A second observation is that the background or timeline of a regulation for the policy area in question does not appear to be decisive for the regulatory coverage. Issues that have been well understood and have given rise to environmental concerns for decades, such as dumped ammunition, still remain essentially unregulated. At the other side of the spectrum new knowledge highlights new regulatory needs. As the chapter on carbon storage illustrates, the concerns surrounding an activity that is so recent that it has not even been put into practice yet has already been subject to detailed regulation at several levels, including laws that specifically concern the Baltic Sea. Hence, more recent environmental concerns are not necessarily more sparingly regulated than older issues. Indeed, the carbon storage example suggests that the swift regulation of a new subject area may provide some regulatory benefits by allowing the key obligations and principles to be settled at international level before states become too politically or economically attached to the regulated activity.

At the same time, those two examples also indicate that it is not necessarily the environmental urgency or need for a given international rule that decides whether it is introduced or not. Factors, such as convenience, political acceptability and remoteness from key state interests may be equally relevant in deciding its regulatory fate. Moreover, the political prioritisation of states is naturally decisive, which explains why eutrophication suddenly became regulated at multiple international layers in the first decade of this millennium while the regulation of historical wrecks and chemical mixtures essentially remain largely untouched, and thus subject to large national variations. The increasingly complex nature of emerging environmental concerns, such as ocean acidification or chemical mixtures, also suggests that technical uncertainties about the form and the nature of the regulation may prevent law-making, irrespective of the political readiness.

The more recent generation of environmental rules adopted by the EU since the introduction of its 'Integrated Maritime Policy'could be a mechanism with which to address such issues in the environmental field. Ecosystem-oriented marine instruments that provide flexibility through generic goals, such as a 'good environmental status'are to be further defined, refined and implemented at national or subnational

level, and in different ways in distinct regions based on their own specific challenges and needs. On the other hand, the chapters have also illustrated how this
uncertainty can spread downwards to national and subnational actors and merely
delegate the responsibilities to a lower level. At international level, too, uncertainty
may be introduced by using the flexibility to introduce tools of unclear legal status.
As noted in the field of eutrophication in particular, the 'soft law'status of the BSAP
and its targets generate uncertainty about the extent of the legal obligations at both
HELCOM and EU level.

The more recent orientation towards eco-system based environmental rules has
not done away with more traditional forms of 'command-and-control'style legislation in individual sectors, whether at EU or international level. While a mix of different regulatory approaches and styles is probably advisable from a policy
perspective (as different types of problems require different solutions), the increasing mix of regulatory styles (general vs. specific, cross-cutting vs. sector-specific,
goal-oriented vs. conduct-oriented, ecosystem-based vs. activity-based and so on)
serves to increase the prospect of normative clashes within and between various
regulatory layers.

10.7   How Are the Gaps Filled?

The gaps presented by unclear or absent rules at international level will normally
have to be addressed by national or subnational rules; the more gaps there are at
international level, the more space there is for such national rules to fill the void.
The chapters in this book have illustrated, for example, how national rules on property rights rules fill the space left by the international salvage law and how the
Finnish Antiquities Act fills the space left by the absence of applicable international
rules on underwater cultural heritage. The chapters have also provided examples of
cases where national rules are prohibited, as is the case regarding coastal state rules
preventing other states from laying pipelines on their continental shelves or imped-
ing the free navigation of foreign ships on their EEZ.

While national rules will ensure that any gaps in international or EU regulation
does not extend across the whole range of applicable regulatory layers, national
solutions are often unsatisfactory from the point of view of the governance of the
Baltic Sea as a whole. There is no certainty that the states will have legislation to
cover the international voids and, even if they do, reliance on national law can lead
to large national differences of regulation and implementation.

In order to reduce the scope for regulatory voids or national divergences, international environmental law has established certain general principles and processes to
be followed in environmental governance. These include the precautionary principle, the polluter pays principle, various obligations relating to conducting the
assessment of environmental impacts and strategic environmental assessments and
to public participation in environmental decision-making as well as more general
obligations for states to cooperate on environmental matters. Many of these

principles are of general applicability throughout the world through the widespread formal acceptance of the conventions in which they appear,[14] and some have been found to represent international customary law, and are thereby binding on all states.[15] At regional level they have been reiterated notably in in the Helsinki Convention and in the Treaty on the Functioning of the European Union and numerous individual EU acts.[16] Yet, it is uncertain how well equipped such general principles are to address the specific issues of environmental governance. At least the authors of the chapters in this book have made very few references to such principles.

An even more recent regulatory development focuses specifically on resolving competing interests between the various uses of and interests in the seas. The concept of marine spatial planning, which is discussed in the chapter by David Langlet, may prove to be of crucial importance when discussing conflicting interests with regard to the oceans in the future. However, as a legal requirement this process currently exists only at the EU level and does not therefore extend to all Baltic Sea states. Moreover, it mainly operates at national level and does not therefore provide a mechanism to balance interests between the different states in the Baltic Sea, even if international and regional coordination is strongly recommended. However, even at national level the rules in their current initial format are excessively vague in terms of substantive content. They focus on procedures and mechanisms and do not include substantive requirements for the plans or rules on the consequences of failing to implement the plans.

10.8 Gaps as Opportunities?

Finally, it should be noted that international regulatory gaps do not only represent problems. It is by no means certain that every issue which gives rise to tension between the states bordering the Baltic Sea needs to be steered by the forceful hand of legislation. Some issues may simply not be important enough to merit regulatory attention, while others can (continue to) be resolved individually by the littoral states at national level. In either case, potential coordination needs may well be

---

[14] For example, the 'precautionary approach', in which lack of scientific certainty shall not be used as a reason to postpone cost-effective measures to prevent environmental degradation, was introduced as Principle 15 in the Rio Declaration in 1992, and has since been reiterated in many international conventions, including the 1992 Convention on Biological Diversity, the 1997 Kyoto Protocol, the 1995 Agreement on Straddling Stocks and Highly Migratory Species and the 1996 London Dumping Protocol.

[15] As regards the precautionary principle, see for example the 2010 judgment of the ICJ in the *Pulp Mills on the River Uruguay Case (Argentina v. Uruguay)* [2010] ICJ Rep 14, para 164 and *Responsibilities and Obligations of States Sponsoring Persons and Entities with Respect to Activities in the Area* (Advisory Opinion) [2011] ITLOS Rep 10, para. 135.

[16] See arts. 11 and 191 of the TFEU; and art. 3 of the 1992 Helsinki Convention

satisfied through lighter guidance by instruments of 'softer' normative force than legal acts.

Even for issues that are deemed serious enough to merit international legislative action, structural or technical problems may stand in the way of effective international rules. Issues may simply be too complex for legislation, or they might require disproportionately costly implementation mechanisms. In such cases, what may seem to be 'gaps' in the regulation may in fact represent the limitations of what legislation can usefully achieve. For those cases, two principal options remain.

Firstly, the understanding of how legal measures operate may require being studied from new perspectives. As the gradual introduction of eco-system based environmental requirements demonstrates, it is feasible to re-think the role and tools of legislation - in the light of new needs without departing from legislation as a tool - by just changing the way it works. Dysfunctional laws call for innovative regulatory solutions to address issues of common concern. There appears to be significant untapped potential in this field for the Baltic Sea area, as is indicated, for example, by the absence of any regulatory measure to date that seeks to address the environmental concerns of the Baltic Sea by means of market-based measures, such as taxation tools, nutrient trading schemes[17] or the traditional approaches that still apply to the enforcement of the rules. However, as the gradual (and slow) shift towards ecosystem-based rules has also illustrated, the process of introducing innovative law and its enforcement is neither quick nor easy and may encounter significant resistance.

Secondly, alternative measures to achieve the desired results may be sought outside the realms of law. Regulatory gaps hence provide an opportunity for other governance mechanisms, such as information-based tools, and economic steering mechanisms such as funding instruments to step in. In this area, the Baltic Sea is already a front-runner, as evidenced by the vast range of institutionalised policy network structures that operate in the region. Most of these structures were established following the collapse of the Soviet Union in the early 1990s to bring together various sector policies and resources, involving a wide range of stakeholders - public, private and civil society - and strengthening international cooperation in various issues of concern to the region. More recently, the EU has been brought in as a key actor in this field too. Its first 'macro-regional strategy' was instituted for the Baltic Sea in 2009, *inter alia,* to support and speed up the implementation of the HELCOM BSAP by means of new governance models. The strategy specifically entails no new legislation, no new budget and no new institutions in support of its implementation.

Regulation and other governance mechanisms are not mutually exclusive approaches. On the contrary, different types of steering mechanism act in parallel and ideally support each other by interacting in the same field within their respective

---

[17] See e.g. S. Hautakangas & M. Ollikainen "Making the Baltic Sea Action Plan workable: a nutrient trading scheme" in Pihlajamäki & Tynkkynen (eds) *Governing the blue-green Baltic Sea - Societal challenges of marine eutrophication prevention (*Finnish Institute of International Affairs, FIIA Report No. 31, 2011) Ch 10. See also http://nutritradebaltic.eu

areas of strength. The inter-relationship of the regulatory and other steering mechanisms deserves more research in general, and the Baltic Sea region provides an ideal test-bed for such research. Some of the challenges linked to the regulatory aspect of that research have been highlighted in the preceding chapters and it is hoped that the on-going BaltReg project, of which this publication forms a part, can help to illuminate the interrelationship between laws and other policy steering mechanisms. It is our belief that a genuine understanding of how this interrelationship works in different substantive areas holds the key to a more effective environmental governance of the Baltic Sea.

Index

© Springer International Publishing AG, part of Springer Nature 2018
H. Ringbom (ed.), *Regulatory Gaps in Baltic Sea Governance*, MARE
Publication Series 18, https://doi.org/10.1007/978-3-319-75070-5